M336
Mathematics and Computing: a third-level course

GROUPS & GEOMETRY

UNIT IB3
FRIEZE PATTERNS

Prepared for the course team by
Peter Strain-Clark

The Open University

This text forms part of an Open University third-level course.
The main printed materials for this course are as follows.

Block 1
Unit IB1	Tilings	
Unit IB2	Groups: properties and examples	
Unit IB3	Frieze patterns	
Unit IB4	Groups: axioms and their consequences	

Block 2
Unit GR1	Properties of the integers
Unit GR2	Abelian and cyclic groups
Unit GE1	Counting with groups
Unit GE2	Periodic and transitive tilings

Block 3
Unit GR3	Decomposition of Abelian groups
Unit GR4	Finite groups 1
Unit GE3	Two-dimensional lattices
Unit GE4	Wallpaper patterns

Block 4
Unit GR5	Sylow's theorems
Unit GR6	Finite groups 2
Unit GE5	Groups and solids in three dimensions
Unit GE6	Three-dimensional lattices and polyhedra

The course was produced by the following team:

Andrew Adamyk (BBC Producer)
David Asche (Author, Software and Video)
Jenny Chalmers (Publishing Editor)
Bob Coates (Author)
Sarah Crompton (Graphic Designer)
David Crowe (Author and Video)
Margaret Crowe (Course Manager)
Alison George (Graphic Artist)
Derek Goldrei (Groups Exercises and Assessment)
Fred Holroyd (Chair, Author, Video and Academic Editor)
Jack Koumi (BBC Producer)
Tim Lister (Geometry Exercises and Assessment)
Roger Lowry (Publishing Editor)
Bob Margolis (Author)
Roy Nelson (Author and Video)
Joe Rooney (Author and Video)
Peter Strain-Clark (Author and Video)
Pip Surgey (BBC Producer)

With valuable assistance from:

Maths Faculty Course Materials Production Unit
Christine Bestavachvili (Video Presenter)
Ian Brodie (Reader)
Andrew Brown (Reader)
Judith Daniels (Video Presenter)
Kathleen Gilmartin (Video Presenter)
Liz Scott (Reader)
Heidi Wilson (Reader)
Robin Wilson (Reader)

The external assessor was:
Norman Biggs (Professor of Mathematics, LSE)

The Open University, Walton Hall, Milton Keynes, MK7 6AA.

First published 1994. Reprinted 2001, 2005.

Copyright © 1994 The Open University

All rights reserved. No part of this publication may be reproduced, stored in a retrieval system or transmitted in any form or by any means, without written permission from the publisher or a licence from the Copyright Licensing Agency Limited. Details of such licences (for reprographic reproduction) may be obtained from the Copyright Licensing Agency Ltd of 90 Tottenham Court Road, London, W1P 9HE.

Edited, designed and typeset by the Open University using the Open University TEX System.

Printed in Malta by Gutenberg Press Limited.

ISBN 0 7492 2161 5

This text forms part of an Open University Third Level Course. If you would like a copy of *Studying with The Open University*, please write to the Central Enquiry Service, PO Box 200, The Open University, Walton Hall, Milton Keynes, MK7 6YZ. If you have not already enrolled on the Course and would like to buy this or other Open University material, please write to Open University Educational Enterprises Ltd, 12 Cofferidge Close, Stony Stratford, Milton Keynes, MK11 1BY, United Kingdom.

CONTENTS

Study guide	4
Introduction	5
1 Describing groups of symmetries	**6**
1.1 Standard form	6
1.2 Groups of symmetries in the plane	11
2 Friezes with reflectional symmetry	**18**
2.1 Type 2	18
2.2 Type 3	22
2.3 Type 4	24
3 Friezes with rotational symmetry	**26**
3.1 Type 5	26
3.2 Type 6	30
3.3 Type 7	31
4 Classifying frieze patterns	**36**
4.1 Isomorphism properties of frieze groups	36
4.2 The frieze group algorithm	38
4.3 Only seven frieze groups	41
4.4 International notation	43
Solutions to the exercises	44
Objectives	54
Index	55

STUDY GUIDE

This unit builds on much of the material in *Units IB1* and *IB2*.

Although the material of this unit is divided into only four sections, you will probably find the study time much the same as for the other units of the course.

The video programme associated with this unit is VC1B *Friezes*, the second programme on Video-cassette 1. The programme is best viewed after you have studied Section 3 but before you study Subsection 4.2, though you *could* view it at any convenient time prior to your study of that subsection.

There is no audio programme associated with this unit.

Your *Geometry Envelope* contains a Frieze Card and two associated overlays. You will need these and the Isometry Toolkit card throughout your study of this unit.

INTRODUCTION

This unit describes in detail the classification of *frieze patterns*, which have existed for centuries as adornments to buildings and other artefacts such as pottery and clothing. Some architectural frieze patterns are shown in Figure 0.1.

You met the concept of a frieze pattern in *Unit IB2*.

Figure 0.1

From a mathematical point of view, frieze patterns may be classified in terms of their groups of symmetries.

Since we shall be dealing with the similarities and differences between several groups, Section 1 begins by showing how some of the example groups we have already met can be represented in a *standard form*.

We then introduce *groups of symmetries* for patterns in the plane. We give both geometric and algebraic descriptions of the symmetries and introduce examples of the seven types of frieze pattern, which will be studied in detail. These seven types of frieze pattern are illustrated on both sides of the Frieze Card in your *Geometry Envelope*.

In Section 2 we deal with the *types of frieze pattern* which have groups of symmetries containing *translations*, *reflections* and *glide reflections* but no rotations. We approach the symmetries both geometrically and algebraically. These contrasting approaches lead to relations between elements and to a complete representation of the groups in standard form.

In Section 3 we deal with the remaining three *types of frieze pattern*, all of which have *rotational* symmetries. As before, the geometry and algebra combine to give us relations between elements and a complete representation of the groups in standard form.

In Section 4 we bring together the results of the previous two sections to obtain an *algorithm* for determining the type of any frieze pattern. Subsequently we prove that the seven types that we have described in detail are the *only* possibilities. We also introduce a notation for frieze groups based on the algorithm and relate this to the *international notation*.

The algorithm is also described informally on the video programme associated with this unit.

1 DESCRIBING GROUPS OF SYMMETRIES

1.1 Standard form

In this unit we shall be looking at the groups of symmetries of all possible types of frieze pattern. You have already met one such group, E_1, in Section 1 of *Unit IB2*. In later units you will be meeting groups of symmetries of lattices and wallpaper patterns; these are exemplified by the group E_2, which you also met in Section 1 of *Unit IB2*.

Later in this unit, when you have met more examples, we shall look in detail at symmetries of frieze patterns in terms of translations, rotations and reflections and the relations between them. However, to begin with, we shall concentrate on some very general points about how small groups can be described and how these descriptions can be combined to give descriptions of larger groups. This work formalizes some of the ideas on describing groups that you met in *Unit IB2*.

The Klein group

One of the smallest groups is the Klein group, V, which you have already encountered in *Unit IB2*. This can be described explicitly as having the four elements

$$\{e, r, h, v\}$$

with the group operation given by:

$$ee = e; \quad er = r; \quad eh = h; \quad ev = v;$$
$$re = r; \quad rr = e; \quad rh = v; \quad rv = h;$$
$$he = h; \quad hr = v; \quad hh = e; \quad hv = r;$$
$$ve = v; \quad vr = h; \quad vh = r; \quad vv = e.$$

Elsewhere (such as in *Unit IB2*) you may have seen this group described as having elements $\{e, a, b, c\}$. We use $\{e, r, h, v\}$ to emphasize that, as we saw in *Unit IB2*, it is isomorphic to $\Gamma(\square)$, the *group of symmetries of the rectangle*.

These relations are often displayed in the form of a Cayley table.

We do not need to specify that $er = r$, etc., as this is true in any group. Moreover, we can express rr as r^2, etc. This gives us the rather shorter description

$$V = \{e, r, h, v : r^2 = e, \ rh = v, \ rv = h,$$
$$hr = v, \ h^2 = e, \ hv = r,$$
$$vr = h, \ vh = r, \ v^2 = e\}.$$

You can quickly convince yourself that an even smaller set of relations is sufficient to enable you to perform all calculations. If we write

$$V^{(1)} = \{e, r, h, v : h^2 = v^2 = e, \ r = hv = vh\},$$

then we can still perform all the same calculations in $V^{(1)}$ as in V, only now using the reduced set of relations. Of course this means that the groups V and $V^{(1)}$ are exactly the same, and so we could use the name V for both. The only point in the superscript $^{(1)}$ is to distinguish the way the group is presented.

You met this small set of relations describing the group $\Gamma(\square)$ in *Unit IB2*.

Example 1.1

Using the relations in the definition of $V^{(1)}$, we can show that $r^2 = e$ as follows:

$$r^2 = rr$$
$$= vhhv \quad \text{(since } r = vh \text{ and } r = hv\text{)}$$
$$= vh^2v$$
$$= vev \quad \text{(since } h^2 = e\text{)}$$
$$= e \quad \text{(since } v^2 = e\text{)}.$$

♦

Exercise 1.1

Use the relations in the definition of $V^{(1)}$ to deduce that $vr = h$.

Using these relations makes it apparent that any one of the symbols r, h and v, say the symbol r, is redundant. The element r could be represented by hv all the time, since we must put it in this form anyway in order to perform calculations. All we need is the description

$$V^{(2)} = \{e, h, v, hv : h^2 = v^2 = e, \ hv = vh\}.$$

There are still four elements in the group $V^{(2)}$, and we can say that they are *expressed in terms of h and v*. By this we mean that each element other than e has been written as an expression involving one or both of the elements h and v: $h = h$, $v = v$ and $r = hv$. Moreover, we have chosen *one particular* expression for each non-identity element of the group; for example, we have chosen hv rather than vh or hv^3 for r. In other words, each element of $V^{(2)}$ is expressed in a **standard form** in terms of the elements h and v, which we call **generators**. The composite of any two of the elements in the group can be reduced to the standard form using only the given relations in the definition of $V^{(2)}$.

We could even express e as h^2, for example, but we do not usually bother as e is automatically an element of every group.

Exercise 1.2

(a) Use the relations in the definition of $V^{(2)}$ to reduce the composite of h and vh to standard form.

(b) What is the inverse of vh, in standard form?

We shall henceforth drop the superscript and refer to $V^{(2)}$ simply as V — the Klein group — since both descriptions represent the same group, having four elements with composition worked out in the same way.

Of course, if we had just been given the description we would have had to make many checks to ensure that it gives a group. However, since we have in the back of our minds the concrete example of the group of symmetries of the rectangle, we need only to ensure that we have given sufficient relations to perform composition.

Finally, we note that our standard form for V can be written in a more generalizable way as follows:

$$V = \{h^m v^n : m, n = 0, 1; \ h^2 = v^2 = e, \ hv = vh\}.$$

We can see that each of the two ways of writing V in standard form are equivalent by taking each of the allowable choices for m and n in turn to give

$$h^0 v^0 = e, \quad h^1 v^0 = h, \quad h^0 v^1 = v, \quad h^1 v^1 = hv.$$

The numerical restrictions

$$m, n = 0, 1$$

on the powers of h and v that appear in the description of the group should be considered as part of the description of the elements. They are different in nature from the relations

$$h^2 = v^2 = e, \quad hv = vh$$

which follow them and which are used to find the composite of two elements in standard form.

We have laboured our treatment of expressing the very small group V in standard form so that we will be able to deal confidently with more substantial examples, such as the following.

The group of symmetries of the regular hexagon

In *Unit IB2* we saw that we can describe the *group of symmetries of the regular hexagon* as

$$D_6 = \{r^m s^n : m = 0, \ldots, 5, \ n = 0, 1; \ r^6 = s^2 = e, \ sr = r^5 s\}.$$

This is very similar to the way that we ended up describing the group V, and is our **standard form** for D_6.

Taking each of the allowable choices for m and n in turn, we obtain the complete set of twelve elements of D_6 explicitly in standard form as

$$\{e, r, r^2, r^3, r^4, r^5, s, rs, r^2 s, r^3 s, r^4 s, r^5 s\}.$$

The relations

$$r^6 = s^2 = e, \quad sr = r^5 s$$

are used to find the composite of two elements in standard form, as the following example and exercise illustrate.

Example 1.2

To find the composite of $r^2 s$ and $r^5 s$, we write

$$r^2 s r^5 s$$

and use the relations to reduce this to the standard form $r^m s^n$, with the powers of r on the left and those of s on the right.

The problem lies in the centre of the expression, where we have an s and a power of r in the wrong order. We need to change sr^5 to something of the form $r^m s^n$. The obvious place to start is the relationship $sr = r^5 s$, but how do we get this to tell us anything about sr^5?

Firstly, since $r^6 = e$, we can write the relation $sr = r^5 s$ as $sr = r^{-1} s$. Then multiplying both sides by s^{-1}, on the right, we can write the relation as a *conjugacy relation*:

$$srs^{-1} = r^{-1}.$$

Raising both sides to the fifth power, we have:

$$(srs^{-1})^5 = (r^{-1})^5$$

$$\Rightarrow \quad srs^{-1} srs^{-1} srs^{-1} srs^{-1} srs^{-1} = r^{-5}$$

$$\Rightarrow \quad sr^5 s^{-1} = r^{-5} \quad \text{(since all the pairs } s^{-1}s \text{ cancel to give } e\text{)}$$

$$= r \quad \text{(since } r^6 = e\text{)}$$

Finally, multiplying both sides by s, on the right, we have

$$sr^5 = rs.$$

So we can solve our original problem. The composite becomes

$$r^2 s r^5 s = r^2 rss \quad \text{(using } sr^5 = rs\text{)}$$

$$= r^3 \quad \text{(using } s^2 = e\text{),}$$

which is in standard form (with $m = 3$, $n = 0$). ♦

You may remember from your previous studies that if x and y are any elements of any group, then xyx^{-1} is said to be conjugate to y. The expression xyx^{-1} is sometimes described as the result of 'conjugating y by x'. We approach our problem of writing $r^2 s r^5 s$ in standard form via conjugation, rather than via repeated use of $sr = r^5 s$, as in Unit IB2, for reasons that will become clear very shortly.

Exercise 1.3

Find the composite of $r^3 s$ and r^4 in standard form.

Now that we have worked through similar details twice, it is time to summarize them as a general result.

> **Theorem 1.1**
>
> If in a group G we have the conjugacy relation
> $$aba^{-1} = c \quad \text{or equivalently} \quad ab = ca,$$
> then, for all $n \in \mathbb{Z}$,
> $$ab^n a^{-1} = c^n \quad \text{or equivalently} \quad ab^n = c^n a.$$

Proof

For $n = 0$, the result is trivially true.

For $n > 0$, the result is proved by combining the conjugacy relation with itself n times, giving
$$\left(aba^{-1}\right)^n = c^n,$$
that is
$$aba^{-1} aba^{-1} \ldots aba^{-1} = c^n.$$
Each adjacent pair $a^{-1}a$ cancels, and so
$$ab^n a^{-1} = c^n.$$

To prove the result for $n < 0$, we first take the inverse of both sides of the conjugacy relation, to obtain
$$\left(aba^{-1}\right)^{-1} = c^{-1},$$
which gives
$$ab^{-1} a^{-1} = c^{-1}.$$
This is a conjugacy relation in G, and we have $-n > 0$. So we can apply the result we have already proved for positive integers to obtain
$$a\left(b^{-1}\right)^{-n} a^{-1} = \left(c^{-1}\right)^{-n},$$
which gives
$$ab^n a^{-1} = c^n.$$
Hence the result holds for all $n \in \mathbb{Z}$. ∎

This innocuous little result will prove to be very powerful in what follows. If we have a relation, such as
$$ab = ca,$$
that enables us to move one element, here a, past another, then we also know how to move it past a power of the other, using the relation
$$ab^n = c^n a.$$

The group of symmetries of a plain rectangular frieze

As our final example in this subsection, we look at the *group, E_1, of symmetries of a plain rectangular frieze*, which you met in *Unit IB2*.

Figure 1.1   A plain rectangular frieze.

The description of E_1 that we obtained in *Unit IB2* is

$$E_1 = \{(t[\mathbf{a}])^n \, x : \ n \in \mathbb{Z}, \ x \in \Gamma(\square); \ h\,t[\mathbf{a}] = t[\mathbf{a}]\,h, \ v\,t[\mathbf{a}] = (t[\mathbf{a}])^{-1}\,v, \ r = hv = vh, \ h^2 = v^2 = e\}.$$

Elements of the group are described in the **standard form**

$$(t[\mathbf{a}])^n \, x,$$

where a symmetry x of the base rectangle is combined with a translation $(t[\mathbf{a}])^n$ through n rectangles, each of length $a = \|\mathbf{a}\|$.

Here, because the group is infinite, we cannot write down individually all the elements. An attempt to do so might start:

$\quad e, r, h, v$ \hfill (choosing $n = 0$)
$\quad t[\mathbf{a}], t[\mathbf{a}]\,r, t[\mathbf{a}]\,h, t[\mathbf{a}]\,v$ \hfill (choosing $n = 1$)
$\quad (t[\mathbf{a}])^{-1}, (t[\mathbf{a}])^{-1}\,r, (t[\mathbf{a}])^{-1}\,h, (t[\mathbf{a}])^{-1}\,v$ \hfill (choosing $n = -1$)
$\quad (t[\mathbf{a}])^2, (t[\mathbf{a}])^2\,r, (t[\mathbf{a}])^2\,h, (t[\mathbf{a}])^2\,v$ \hfill (choosing $n = 2$)

The relations in the definition of E_1 tell us how to compose elements. The first two tell us how translations can be moved past reflections, and the rest just relate to the products of elements in $\Gamma(\square)$. We can therefore simplify the description to

$$E_1 = \{(t[\mathbf{a}])^n \, x : \ n \in \mathbb{Z}, \ x \in \Gamma(\square); \ h\,t[\mathbf{a}] = t[\mathbf{a}]\,h, \ v\,t[\mathbf{a}] = (t[\mathbf{a}])^{-1}\,v\},$$

where it is understood that all the usual relations in $\Gamma(\square)$ still apply.

This simplified description was hinted at in Unit IB2.

What appears to be missing from this description is any mention of how translations compose. We can overcome this omission by defining the set

$$T_1 = \{(t[\mathbf{a}])^n : \ n \in \mathbb{Z}\}.$$

The definitions of $t[\mathbf{a}]$ and $(t[\mathbf{a}])^n$ immediately tell us that T_1 is a *group*. It is closed since

$$(t[\mathbf{a}])^m \, (t[\mathbf{a}])^n = (t[\mathbf{a}])^{m+n}$$

and $m + n \in \mathbb{Z}$. Associativity clearly holds. The identity element is

$$(t[\mathbf{a}])^0 = e,$$

and the inverse of every element is in T_1 since

$$(t[\mathbf{a}]^n)^{-1} = (t[\mathbf{a}])^{-n}$$

and $-n \in \mathbb{Z}$.

Furthermore, by the definition of $t[\mathbf{a}]$, we have a *distinct* translation $(t[\mathbf{a}])^n$ for each $n \in \mathbb{Z}$, so that:

$(t[\mathbf{a}])^n = e \quad$ if and only if $\quad n = 0;$
$(t[\mathbf{a}])^m = (t[\mathbf{a}])^n \quad$ if and only if $\quad m = n.$

In other words, no relation between elements of T_1 exists that is capable of simplifying any composite element $(t[\mathbf{a}])^m \, (t[\mathbf{a}])^n$ other than to write it as $(t[\mathbf{a}])^{m+n}$. Therefore

$$T_1 = \{(t[a])^n : \ n \in \mathbb{Z}\}$$

is a complete description of the group of translations of a plain rectangular frieze, and does not require any additional relations to be added to it to enable us to compose any two elements of the group and write the result in the standard from $(t[\mathbf{a}])^n$.

In light of this, we can make our final simplification to the description of the group of symmetries of a plain rectangular frieze as

$$E_1 = \{xy : \ x \in T_1, \ y \in V; \ ht = th, \ vt = t^{-1}v\},$$

where it is understood that all the usual relations in T_1 and V still apply.

We use x to stand for a *general* element of T_1, and y for a *general* element of V, so that the symbol t can be used for the *particular* element $t[\mathbf{a}]$ of T_1.

The translation $t[\mathbf{a}]$ plays a fundamental role in E_1, and in all the frieze groups we shall describe, so we shall shorten the notation and write $t[\mathbf{a}]$ as just t whenever convenient. Similarly $\Gamma(\square)$ and its subgroups play a vital part, and so we shall generally use the shorter (Klein group) notation V instead of $\Gamma(\square)$.

and r, h and v can be used for the *particular* elements of V that you have already seen.

Example 1.3

To put the composite rt^2 in standard form, we move the translation to the left past r, and to do this we use the definition of r in terms of the generators h and v of V. We obtain

$$\begin{aligned}rt^2 &= vht^2 &&\text{(since } r = vh \text{ in } V\text{)}\\ &= vt^2h &&\text{(since } ht = th \text{ and hence, by Theorem 1.1, } ht^2 = t^2h\text{)}\\ &= t^{-2}vh &&\text{(since } vt = t^{-1}v \text{ and hence, by Theorem 1.1, } vt^2 = t^{-2}v\text{)}\\ &= t^{-2}r &&\text{(since } r = vh \text{ in } V\text{)},\end{aligned}$$

which is in standard form. ♦

1.2 Groups of symmetries in the plane

In what follows we shall be looking at the symmetry groups of many patterns, and consequently we need to introduce a systematic notation. We shall be considering only two-dimensional patterns, and so we begin with some generalities about the two-dimensional plane, $\mathbb{R}^2$, in which they are all assumed to reside.

Isometries of the plane

We have seen in *Unit IB1* that the *plane isometries* (i.e. isometries of the plane, $\mathbb{R}^2$) consist of all translations, rotations, reflections and glide reflections. We have also seen that, once an origin and a coordinate system have been chosen, any plane isometry can be written in the *standard form*

$$t[\mathbf{p}]\,\lambda[\mathbf{A}]$$

where $\mathbf{p}$ is a vector and $\mathbf{A}$ is an orthogonal matrix. If

$$\mathbf{p} = \begin{bmatrix} p \\ q \end{bmatrix} \quad \text{and} \quad \mathbf{A} = \begin{bmatrix} a & c \\ b & d \end{bmatrix},$$

then this can be written in *explicit form* as the function

$$\begin{bmatrix} x \\ y \end{bmatrix} \mapsto \begin{bmatrix} a & c \\ b & d \end{bmatrix}\begin{bmatrix} x \\ y \end{bmatrix} + \begin{bmatrix} p \\ q \end{bmatrix},$$

or (more compactly) as

$$(x, y) \mapsto (ax + cy + p, bx + dy + q).$$

In Section 5 of *Unit IB1*, we introduced the Isometry Toolkit, for manipulating isometries, which you will find in your *Geometry Envelope* and which you should have at hand from now on when studying this unit.

Recall that rotation through an angle θ, written $r[\theta]$, and reflection in a line making an angle θ with the x-axis, written $q[\theta]$, are represented by the following matrices:

$$r[\theta] \quad \text{by} \quad \begin{bmatrix} \cos\theta & -\sin\theta \\ \sin\theta & \cos\theta \end{bmatrix};$$

$$q[\theta] \quad \text{by} \quad \begin{bmatrix} \cos 2\theta & \sin 2\theta \\ \sin 2\theta & -\cos 2\theta \end{bmatrix}.$$

When there is no possible confusion, we shall not distinguish between the origin-preserving isometry $\lambda[\mathbf{A}]$ and the matrix $\mathbf{A}$.

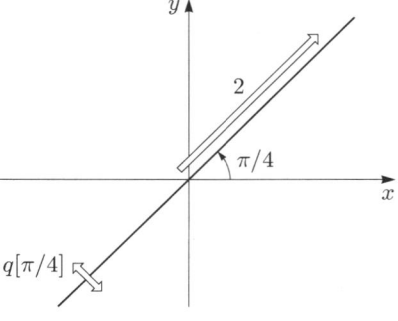

Figure 1.2

Exercise 1.4

Write in explicit form the glide reflection which first reflects in the line making an angle $\pi/4$ with the x-axis and then translates a distance 2 along that line (see Figure 1.2).

Hint You may find the reverse side of the Isometry Toolkit useful.

The set of all plane isometries is, in fact, a *group*. The closure and inverses axioms follow directly from Theorem 3.1 of *Unit IB1*. The identity element is the identity isometry $t[\mathbf{0}]\,\lambda[\mathbf{I}]$. The associativity axiom follows as a consequence of the rule for composing affine transformations (Rule 4 of *Unit IB1*). This group is denoted by

$$\Gamma.$$

Γ is the Greek capital letter *Gamma*.

The group Γ of isometries of the plane has two important subgroups to which we often want to refer. Firstly, there is the subgroup consisting only of *direct* isometries, i.e. translations and rotations. We refer to this subgroup as

$$\Gamma^+.$$

Secondly, there is the subgroup of translations, which we refer to as

$$\Delta.$$

Δ is the Greek capital letter *Delta*.

This group Δ is a subgroup of both Γ^+ and Γ, and we have the hierarchy of inclusions

$$\Gamma \supseteq \Gamma^+ \supseteq \Delta.$$

We can use the Isometry Toolkit to deduce that Γ^+ and Δ are sub*groups* of Γ.

The notation $A \supseteq B$ means that the set A *includes* the set B, i.e. that B is a subset of A; thus $A \supseteq B$ and $B \subseteq A$ are equivalent.

Symmetry groups of figures and patterns

In *Unit IB1* we developed the algebra of isometries in order to study the *symmetries* of the figures and patterns that can be drawn in the plane — a symmetry being a plane isometry that maps the figure or pattern exactly onto itself.

At this point, we need to be clear about exactly what we mean by 'mapping a figure or pattern exactly onto itself'.

The plain rectangular frieze, which you studied in Subsection 1.1, is a pattern which lies within a subset of $\mathbb{R}^2$ consisting of a horizontal strip. *Every* horizontal translation, not only those by integer multiples of $\mathbf{a}$, will map this *strip* to itself. But we do not count $t[\sqrt{2}\mathbf{a}]$, for example, as a symmetry of the *frieze*. This is because, after moving the frieze by $\sqrt{2}\mathbf{a}$, we can tell at a glance that it has moved — which we could *not* do if we had moved it by an *integer* multiple of $\mathbf{a}$. The reason for this is that the *inked part* of the frieze — the set of lines forming the rectangle boundaries — is *not* mapped onto itself by $t[\sqrt{2}\mathbf{a}]$ but *is* mapped to itself by $t[n\mathbf{a}]$ for any integer n.

Similarly, although a tiling $\mathcal{T}$ of the plane covers the whole plane, by definition, a symmetry of $\mathcal{T}$ is not just any plane isometry, but is one that maps the set of tile boundaries, i.e. the net of the tiling, exactly onto itself.

In each case, therefore, we have a subset of the plane (the set which we ink when we make a drawing), and it is *this subset* which has to be mapped to itself in order that an isometry may be regarded as a symmetry of a figure or pattern. Thus we arrive at the following definitions.

Definition 1.1 Plane figure

A **plane figure** is a subset of $\mathbb{R}^2$, such as the inked subset in a drawing of a frieze, tiling or other pattern.

Definition 1.2 Symmetry and symmetry group

Let P be a plane figure. A **symmetry** of P is a plane isometry f such that $f(P) = P$. The **symmetry group** of P is the group of all symmetries of P, and is denoted by $\Gamma(P)$.

In the special case when $P = \mathbb{R}^2$, then the group of all symmetries of the plane is just the group of all plane isometries, i.e. $\Gamma(\mathbb{R}^2) = \Gamma$.

As in the case of Γ on page 12, we can use our results from *Unit IB1* to show that $\Gamma(P)$ is indeed a group, for any plane figure P.

Also, just as the group Γ has subgroups Γ^+ and Δ, so the symmetry group of a pattern P has similar subgroups.

Definition 1.3 Direct symmetry group and translation group

Let P be a plane figure. Then its **direct symmetry group**, $\Gamma^+(P)$, is the subgroup of $\Gamma(P)$ consisting of direct isometries, while its **translation group**, $\Delta(P)$, is the subgroup of $\Gamma(P)$ consisting of translations. The elements of $\Gamma^+(P)$ are known as **direct symmetries** and the elements of $\Delta(P)$ are known as **translational symmetries**.

We have the obvious inclusions

$$\Gamma(P) \supseteq \Gamma^+(P) \supseteq \Delta(P).$$

In the special case when $P = \mathbb{R}^2$, then the direct symmetry group of the plane is just the group of all direct isometries, i.e. $\Gamma^+\left(\mathbb{R}^2\right) = \Gamma^+$.

Similarly, when $P = \mathbb{R}^2$, $\Delta(P) = \Delta\left(\mathbb{R}^2\right) = \Delta$.

Example 1.4

Let R be the subset of $\mathbb{R}^2$ consisting of a rectangle centered at the origin of the plane and with sides parallel to the coordinate axes, as in Figure 1.3.

In this case, the boundary of the rectangle has the same symmetry group as the whole rectangle, so it does not matter whether we ink just the boundary, getting a white rectangle, or the interior as well, getting a black rectangle!

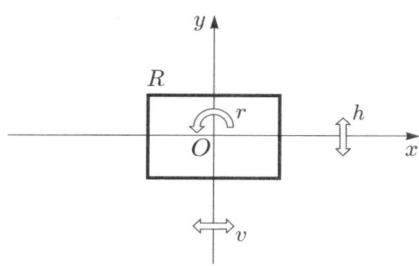

Figure 1.3

We shall describe, explicitly, the subgroups and inclusions

$$\Gamma(R) \supseteq \Gamma^+(R) \supseteq \Delta(R).$$

The group of symmetries of the rectangle R is

$$\Gamma(\square) = V = \{e, r, v, h\}$$

with the relations we have seen earlier.

Since each symmetry of R preserves the origin, they are orthogonal transformations, represented by the corresponding orthogonal matrices. We have:

See Subsection 5.1 of *Unit IB1*.

e, which is represented by $\begin{bmatrix} 1 & 0 \\ 0 & 1 \end{bmatrix}$;

the rotation

$r = r[\pi]$, which is represented by $\begin{bmatrix} -1 & 0 \\ 0 & -1 \end{bmatrix}$;

the reflection

$h = q[0]$, which is represented by $\begin{bmatrix} 1 & 0 \\ 0 & -1 \end{bmatrix}$;

and the reflection

$v = q[\pi/2]$, which is represented by $\begin{bmatrix} -1 & 0 \\ 0 & 1 \end{bmatrix}$.

The subgroup of direct symmetries is
$$\Gamma^+(R) = \{e, r: r^2 = e\}.$$
The subgroup of translations
$$\Delta(R) = \{e\}$$
contains only the identity transformation e. ◆

Exercise 1.5

Let O be the 'figure' consisting of only the origin of the plane. Describe the subgroups and inclusions
$$\Gamma(O) \supseteq \Gamma^+(O) \supseteq \Delta(O).$$

Symmetries of a frieze

Let F be the subset of $\mathbb{R}^2$ consisting of the inked part of a plain rectangular frieze with one of its rectangles, R, centred at the origin O of the plane and with sides parallel to the coordinate axes, as shown in Figure 1.4.

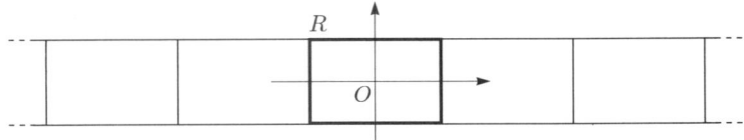

Figure 1.4

F can be thought of as the union of the boundaries of an infinite set of rectangles placed side by side. Equivalently, F can be thought of as consisting of two infinite horizontal straight lines with an infinite number of evenly spaced vertical line segments running between them.

We have seen that the symmetries of F are translations, rotations, reflections and glide reflections, and form the group

$$\Gamma(F) = E_1 = \{xy: x \in T_1, y \in V; ht = th, vt = t^{-1}v\}.$$

See *Unit IB2* and Subsection 1.1 of this unit.

When we consider only the direct symmetries, we must discard reflections and glide reflections and leave only the translations and rotations, giving

$$\Gamma^+(F) = \{x, xr: x \in T_1; r^2 = e, rt = t^{-1}r\}.$$

Finally, when we look for the subgroup of translations, we are left with

$$\Delta(F) = T_1 = \{t^n: n \in \mathbb{Z}\}.$$

Soon we shall see that it is the fact that $\Delta(F)$ is of this form which we take as characterizing a frieze.

When describing $\Gamma^+(F)$, we have had to go back to describing the elements explicitly as x or xr and including the relationship $r^2 = e$, because we have no notation for the subgroup of V containing just the elements e and r. Since we shall frequently meet this and other subgroups of V, we now introduce some notation for them. The subgroups of V are:

We also had to replace the relations $ht = th$ and $vt = t^{-1}v$ by the relation $rt = t^{-1}r$ which explicitly links r and t. See *Unit IB2*, in particular Exercise 1.10(a).

(a) the whole group V;

(b) the subgroup of order 2 generated by the rotation r:
$$R_r = \{e, r: r^2 = e\};$$

(c) the subgroups of order 2 generated by a single reflection:
$$Q_h = \{e, h: h^2 = e\};$$
$$Q_v = \{e, v: v^2 = e\};$$

(d) the trivial subgroup $\{e\}$.

Note that the three subgroups R_r, Q_h, Q_v are all isomorphic as abstract groups, i.e. *algebraically* they are essentially the same. However, R_r consists of direct symmetries and is thus *geometrically* different from the groups Q_h and Q_v, each of which contains a reflection.

Isomorphisms are defined formally in *Unit IB4*.

Using this notation for subgroups we can now write

$$\Gamma^+(F) = \{xy : x \in T_1, y \in R_r; rt = t^{-1}r\},$$

where it is understood that all the usual relations in T_1 and R_r still apply.

Exercise 1.6

Write down the symmetry groups of the patterns, based on rectangles, shown in Figure 1.5.

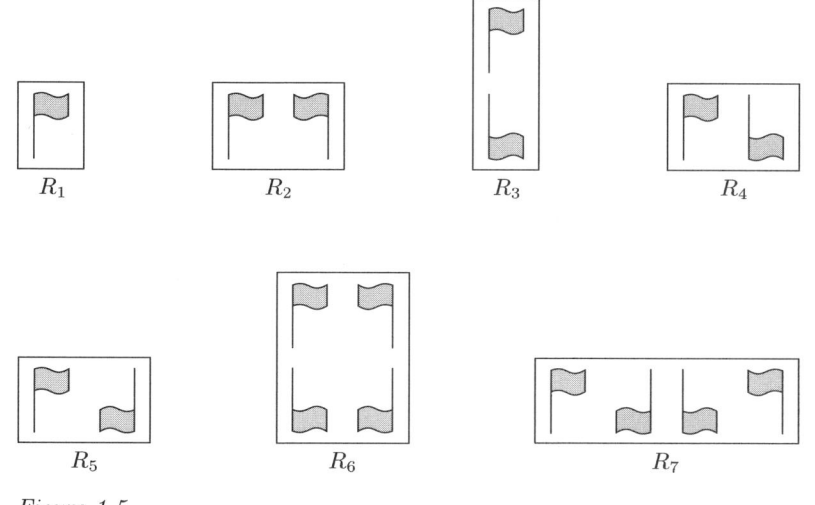

Figure 1.5

It is the interplay between the symmetry groups of the seven rectangular patterns in Figure 1.5 and the group of translations T_1 which accounts for the variety of friezes that we shall study next. But first we need a formal definition of a frieze.

Definition 1.4 Frieze (pattern)

A **frieze** or **frieze pattern** F is a plane figure bounded by two infinite parallel lines and which has T_1 as its group of translations. That is,

$$\Delta(F) = T_1,$$

where this group is generated by a single translation $t = t[\mathbf{a}]$ through a suitable minimum non-zero distance $a = ||\mathbf{a}||$ in the direction of the parallel lines.

Exercise 1.7

Which of the plane figures in Figure 1.6 is a frieze?

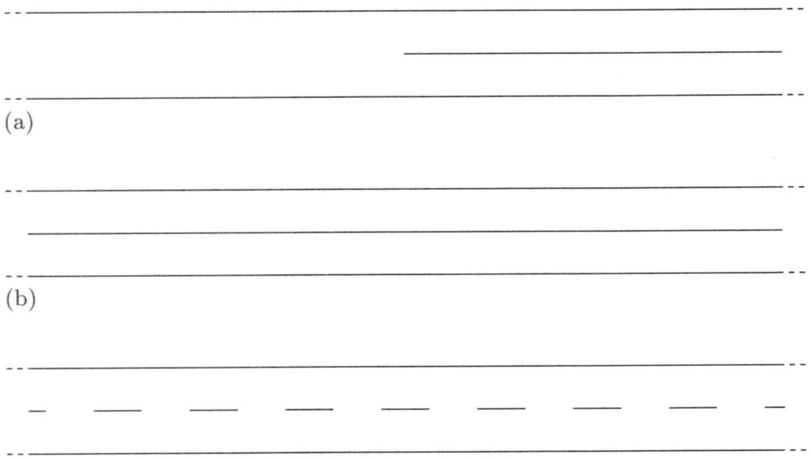

```
-- _____   --
      _____
```
(a)

```
-- _____   --
      _____
-- _____   --
```
(b)

```
-- _____   --
  _   ___   ___   ___   ___   ___   ___   _
-- _____   --
```
(c)

Figure 1.6

Standard examples of the seven types of frieze pattern, corresponding to the seven patterns in Figure 1.5, are shown on the Frieze Card in your *Geometry Envelope*, and on the overlays that go with it (one for each side). We shall refer to them as $F_1, \ldots, F_7$, and, until we establish the International Notation in Subsection 4.4, we shall say that they are of *Type 1, ..., Type 7* respectively. We shall investigate their symmetry groups, known as **frieze groups**, in turn.

We shall postpone, until Subsection 4.3, a proof of the fact that these are the only seven types of frieze possible.

If you look at the Frieze Card, you will see that a flag motif is used on each frieze. The reason for this is that a single flag with a flagpole has no symmetries apart from the trivial symmetry. This makes it easier to sort out the symmetries of the frieze as a whole.

We now ask you to examine for yourself the symmetry group of the standard Type 1 frieze, F_1, shown on the Frieze Card and in Figure 1.7.

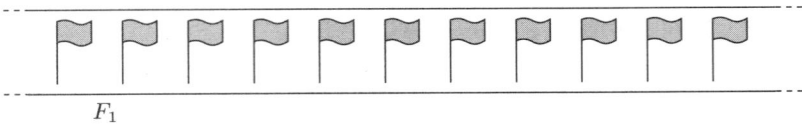

F_1

Figure 1.7

Exercise 1.8

Use the Frieze Card and the Overlay for Side 1 of that card to determine the symmetries of the frieze F_1. Deduce the symmetry groups $\Gamma(F_1)$, $\Gamma^+(F_1)$ and $\Delta(F_1)$.

You probably found, in using the Frieze Card and the Overlay for Side 1, that most of the possible symmetries of $\mathbb{R}^2$ could be disregarded straightaway because the horizontal strip, on which the flags are drawn, did not map to itself. In fact, this observation can be extended if we define the **centre line** of a frieze to be the infinite line parallel to and exactly halfway between the two parallel lines bounding the frieze; then we can see that any symmetry of the frieze must fix the centre line as a whole. This principle will prove very useful, when we come to examine more complicated friezes, as it allows us (in the following theorem) to use the geometric classification of isometries from Section 5 of *Unit IB1* to narrow down very considerably the set of *possible* symmetries of a frieze.

> ### Theorem 1.2 Possible symmetries of a frieze
>
> The only possible symmetries of a frieze are:
> - the identity;
> - translations parallel to the centre line;
> - rotations through π about points lying on the centre line;
> - reflections in lines perpendicular to the centre line;
> - the reflection in the centre line;
> - glide reflections with the centre line as axis.

Proof

We have observed that any symmetry of a frieze must fix the centre line as a whole. Bearing this in mind, let us go through the six geometric types of isometry given in Theorem 5.1 of *Unit IB1*.

(a) *The identity.* Yes, clearly this is always a symmetry of any frieze!

(b) *Non-zero translation.* This fixes lines parallel to the direction of translation (and no other). Thus, translations that are symmetries of the frieze must be parallel to the centre line.

(c) *Rotation through an angle other than 0 or π.* This fixes no lines, and so cannot be a symmetry of any frieze.

(d) *Rotation through π about a point C.* This fixes lines through C (and no others). Hence, for such a rotation to be a symmetry of the frieze, C must be on the centre line.

(e) *Reflection in a line L.* The only lines fixed by such a reflection are L itself and lines perpendicular to L. Hence, for the centre line to be fixed, the axis of reflection must be the centre line itself or must be perpendicular to the centre line.

(f) *Glide reflection.* The only line fixed by a glide reflection is the glide reflection axis. Thus the only glide reflections that can be symmetries of a frieze are those having the centre line as axis. ∎

In Exercise 1.8 we looked at a type of frieze that exhibits *only* translational symmetries, which we refer to as a Type 1 frieze. In the next two sections we shall look at types of friezes which also exhibit reflection symmetries and/or rotational symmetries.

2 FRIEZES WITH REFLECTIONAL SYMMETRY

In this section we deal with friezes which exhibit indirect symmetries but no rotational symmetries.

2.1 Type 2

The geometry of reflections

Consider the symmetries of our standard Type 2 frieze, F_2, shown in Figure 2.1.

Have the Frieze Card and the Overlay for Side 1 handy and use them to help you to follow any of our arguments which are not immediately obvious to you.

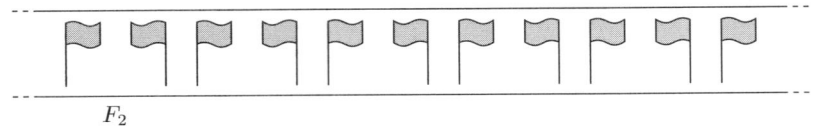

Figure 2.1

The group of translational symmetries is

$$\Delta(F_2) = T_1,$$

where the generating translation t moves each flag two places to the right, as shown in Figure 2.2.

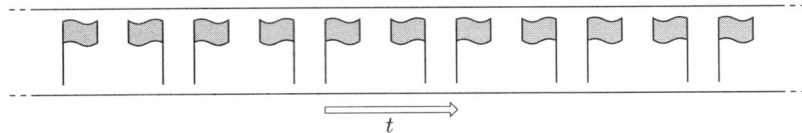

Figure 2.2

There are no rotational symmetries, since rotating the frieze through a half-turn about any point on the centre line will turn the flags upside-down. Hence the translations are the only direct symmetries, and

$$\Gamma^+(F_2) = \Delta(F_2) = T_1.$$

To determine the other symmetries of F_2, we first note that, since the generating translation t moves each flag *two* places to the right, we can consider the translations as moving blocks of *two* flags, such as in Figure 2.3. The block of flags in Figure 2.3 forms a *base rectangle* for the frieze, in that the frieze can be formed by placing copies of the base rectangle side by side. This base rectangle is identical to the rectangular pattern R_2 which we studied in Exercise 1.6. There we saw that its group of symmetries is Q_v, which is generated by the reflection v in its vertical axis of symmetry, as Figure 2.4 illustrates.

Note that we choose as base rectangle the *smallest* rectangle that can be used to build up the frieze. An alternative choice of base rectangle for F_2 is shown below. Either choice would result in the same description for $\Gamma(F_2)$.

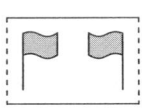

Figure 2.3

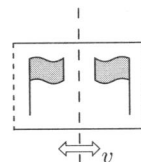

Figure 2.4

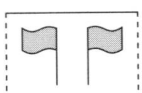

If we now consider the entire frieze, we see that v is indeed a symmetry of it, as Figure 2.5 illustrates.

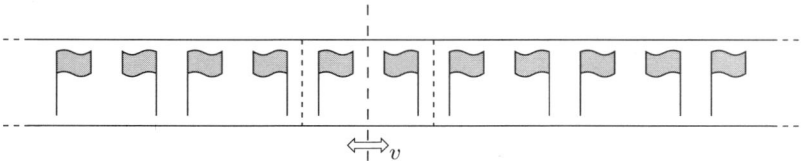

Figure 2.5

Since the composite of two symmetries is again a symmetry, we can take this reflection v, along with all the translations, and form the following set of symmetries of F_2:

$$\{xy:\ x \in T_1,\ y \in Q_v;\ vt = t^{-1}v\}.$$

The relation $vt = t^{-1}v$ was used in our description of E_2 in Subsection 1.1. You can check that this relation holds using the Frieze Card and the Overlay for Side 1. For the composite vt, you first translate one step to the right and then reflect in the vertical axis of symmetry of the base rectangle. The same effect is obtained by performing the composite $t^{-1}v$, where you first reflect in the vertical axis of symmetry of the base rectangle and then translate to the left.

You checked the same relation in *Unit IB2*, Exercise 1.10(c).

You can also derive this relation from Equation 6b of the Isometry Toolkit, by choosing a coordinate system with x-axis the centre line and with the origin lying on the vertical axis of symmetry of the base rectangle. Then, with $t = t[\mathbf{a}]$ and $v = q[\pi/2]$, and since $q[\pi/2](\mathbf{a}) = -\mathbf{a}$, we have:

$$\begin{aligned}
vt &= q[\pi/2]\, t[\mathbf{a}] \\
&= t[q[\pi/2](\mathbf{a})]\, q[\pi/2] \\
&= t[-\mathbf{a}]\, q[\pi/2] \\
&= t^{-1}v.
\end{aligned}$$

Now, as you may have realized, the above set is a group. We can use the relation $vt = t^{-1}v$ and Theorem 1.1 to show that it is closed and that the inverses axiom holds. Associativity follows from the associativity of isometries. The identity axiom holds since $e \in T_1$ and $e \in Q_v$. Therefore, we certainly know that we have a subgroup of the group of symmetries of F_2. That is,

$$\Gamma(F_2) \supseteq \{xy:\ x \in T_1,\ y \in Q_v;\ vt = t^{-1}v\}.$$

It remains to determine whether there are any further symmetries of F_2.

Exercise 2.1

Write down, in terms of t and v, all the symmetries in the above subgroup.

We have considered the reflection symmetry v which reflects the frieze in the vertical axis of symmetry of our base rectangle. But any translate of this rectangle might easily have been chosen as the base rectangle, and so there are reflection symmetries in their vertical axes of symmetry. We shall temporarily write v_n for the reflection symmetry derived from the rectangle translated n steps to the right ($n \in \mathbb{Z}$). Explicitly this gives symmetries

A *translate* of part of a frieze F is a copy of that part obtained by translating it by an element of $\Delta(F)$, the group of translational symmetries of the frieze.

$$\ldots,\ v_{-2},\ v_{-1},\ v_0,\ v_1,\ v_2,\ \ldots$$

where v_0 is the basic reflection v, v_{-1} is the reflection in the vertical axis of symmetry of the rectangle translated one step to the left, etc. — as Figure 2.6 illustrates.

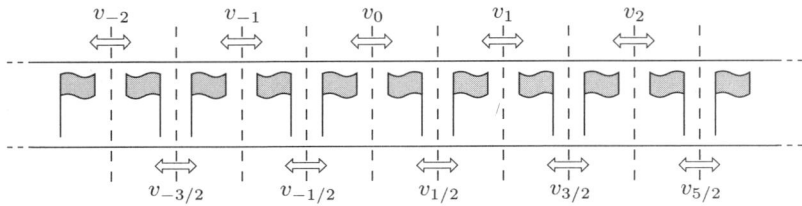

Figure 2.6

Further inspection of the frieze shows that there are also reflection symmetries in the vertical axes separating translates of our original base rectangle. These axes are found by translating the axis of symmetry of the base rectangle through an integer number of steps and then a further half step to the right. We shall temporarily write $v_{n+1/2}$ for the reflection symmetry derived by translating the axis n steps ($n \in \mathbb{Z}$) and then a further half step to the right. Explicitly this gives symmetries

$$\ldots, v_{-3/2}, v_{-1/2}, v_{1/2}, v_{3/2}, v_{5/2}, \ldots$$

where, for example, $v_{1/2}$ and $v_{-1/2}$ are reflections in the vertical axes forming the right and left sides of the base rectangle (see Figure 2.6 again).

Exercise 2.2

What is the composite symmetry

$$v_{-1/2} v_{1/2}?$$

You may find it useful to use the Frieze Card and the Overlay for Side 1. Alternatively, you may prefer to use the Isometry Toolkit: choose a coordinate system, and begin by using Equation 12 of the Toolkit.

This systematic approach is based on the second, alternative solution to Exercise 2.2 (using the Isometry Toolkit).

The algebra of reflections

In order to treat reflections systematically, we now set up a coordinate system with its origin at the centre of the base rectangle and the x-axis pointing horizontally to the right along the frieze, as shown in Figure 2.7.

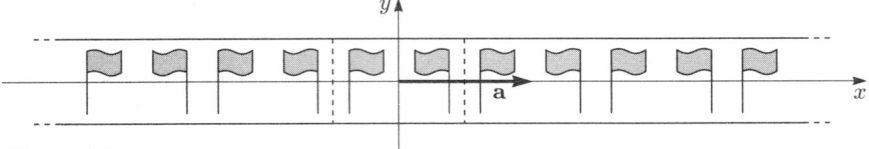

Figure 2.7

Thus, for each integer n, the reflection v_n is a reflection in the line inclined at $\pi/2$ to the x-axis and passing through the point $n\mathbf{a}$ (where $t = t[\mathbf{a}]$). In the notation of the Isometry Toolkit,

$$v_n = q[n\mathbf{a}, \pi/2]. \tag{2.1}$$

Similarly, the reflection $v_{n+1/2}$ is a reflection in the line inclined at $\pi/2$ to the x-axis and passing through the point $\left(n + \frac{1}{2}\right)\mathbf{a}$. In the notation of the Isometry Toolkit,

$$v_{n+1/2} = q\left[\left(n + \tfrac{1}{2}\right)\mathbf{a}, \pi/2\right]. \tag{2.2}$$

Now any multiple of $\mathbf{a}$ is perpendicular to any of these axes of reflection, so we can use Equation 12 of the Isometry Toolkit to re-express Equations 2.1 and 2.2 as follows:

$$\begin{aligned}v_n &= t[2n\mathbf{a}]\, q[\pi/2] \\ &= t^{2n} v \quad \text{(since } t = t[\mathbf{a}] \text{ and } v = q[\pi/2]\text{);}\end{aligned} \tag{2.3}$$

$$\begin{aligned}v_{n+1/2} &= t[(2n+1)\mathbf{a}]\, q[\pi/2] \\ &= t^{2n+1} v.\end{aligned} \tag{2.4}$$

That is to say, v_n can be achieved by following a vertical reflection through the origin by a translation $2n$ steps to the right, whereas $v_{n+1/2}$ can be achieved by following a vertical reflection through the origin by a translation $2n+1$ steps to the right. In either case, we have the following geometric principle.

> *Perpendicular translation principle*
>
> If a reflection is followed by a translation perpendicular to the reflection line, then the resulting isometry is a reflection with axis moved *half* the distance of the translation.

This principle is, of course, implicit in Equation 12 of the Isometry Toolkit. If we put $\mathbf{b} = 2\mathbf{c}$ in this equation and exchange the left-hand and right-hand sides, it reads

$$t[\mathbf{b}]\, q[\theta] = q\big[\tfrac{1}{2}\mathbf{b}, \theta\big],$$

which is a restatement of the principle.

We shall now consider the explicit forms of these symmetries, assuming that the vector $\mathbf{a}$ is $(a, 0)$.

Example 2.1
The symmetry t^{-1} moves every point a distance a to the left. Thus,

$$t^{-1} : (x, y) \mapsto (x - a, y).$$

♦

Exercise 2.3 _____

Write out the explicit form of t^n ($n \in \mathbb{Z}$).

Example 2.2
Consider the symmetry v_n. From Equation 2.3 we have

$$\begin{aligned} v_n &= t[2n\mathbf{a}]\, q[\pi/2] \\ &= t[(2na, 0)]\, q[\pi/2]. \end{aligned}$$

Using Equation 23 of the Isometry Toolkit, and noting that $\cos 2(\pi/2) = -1$ and $\sin 2(\pi/2) = 0$, we obtain

$$v_n : (x, y) \mapsto (2na - x, y),$$

or, alternatively, using Equation 23a of the Toolkit,

$$v_n : \begin{bmatrix} x \\ y \end{bmatrix} \mapsto \begin{bmatrix} -1 & 0 \\ 0 & 1 \end{bmatrix} \begin{bmatrix} x \\ y \end{bmatrix} + \begin{bmatrix} 2na \\ 0 \end{bmatrix}.$$

♦

Exercise 2.4 _____

Write out the explicit form of $v_{n+1/2}$.

Finally, we leave it to you to show that the symmetries t^n, v_n and $v_{n+1/2}$, for all $n \in \mathbb{Z}$, constitute the group which we met earlier.

Exercise 2.5 _____

Show that the set $\{t^n, v_n, v_{n+1/2} : n \in \mathbb{Z}\}$ is exactly the group

$$\{xy : x \in T_1,\ y \in Q_v;\ vt = t^{-1}v\}$$

which was considered earlier in this subsection.

The symmetry group in standard form

We have now shown that the group of symmetries

$$\{xy : x \in T_1,\ y \in Q_v;\ vt = t^{-1}v\}$$

constitutes all the translational symmetries, and all reflection symmetries of the form v_n or $v_{n+1/2}$, of F_2. Is this the whole symmetry group of F_2?

The answer is *yes*. The group T_1 is by definition the translation group. We looked for all possible vertical axes of reflection, finding only two different types corresponding to the reflections v_n $(n \in \mathbb{Z})$ and $v_{n+1/2}$ $(n \in \mathbb{Z})$. By Theorem 1.2, the only other possibilities are rotations, the reflection in the centre line, and glide reflections with the centre line as glide reflection axis. However, all of these turn the flags upside-down, and there are no upside-down flags for them to map onto! Therefore there *cannot* be any such symmetries, and we *must* have accounted for the whole symmetry group. Thus we now know that

$$\Gamma(F_2) = \{xy : \ x \in T_1, \ y \in Q_v; \ vt = t^{-1}v\}.$$

The expression xy (where $x \in T_1$ and $y \in Q_v$, so that $x = t^n$ and y is either e or v) is the **standard form** of an element of $\Gamma(F_2)$. Thus a translation t^n is automatically in standard form, and Equations 2.3 and 2.4 show us how to express in standard form the reflections which we temporarily denoted by v_n and $v_{n+1/2}$.

> Strictly speaking, the standard form of t^n should be $t^n e$, but we normally omit the symbol e in these contexts. Similarly we write v rather than ev for v_0.

Exercise 2.6

Express in standard form the result of performing the translation t^n followed by the reflection v_i.

Exercise 2.7

Express in standard form the result of performing the reflection v_i followed by the reflection v_j.

2.2 Type 3

Consider the symmetries of our standard Type 3 frieze, F_3, shown in Figure 2.8.

> You may find it helpful to continue to use the Frieze Card and the Overlay for Side 1.

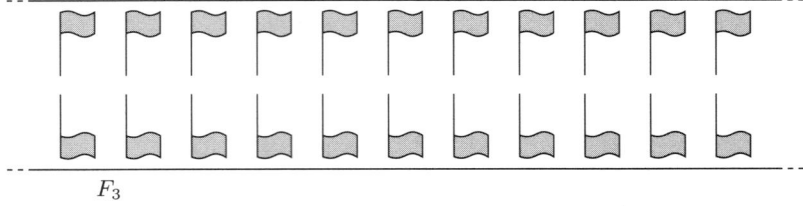

Figure 2.8

For this frieze the flags come in pairs, one flag of a pair near the top of the frieze and one near the bottom. Once again the group of translational symmetries is

$$\Delta(F_3) = T_1,$$

where the generating translation t moves each pair of flags one place to the right, as shown in Figure 2.9.

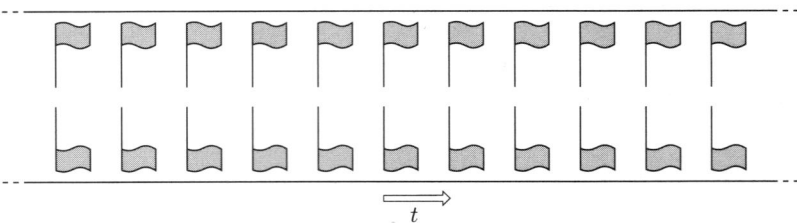

Figure 2.9

There are no rotational symmetries, and so the group of direct symmetries is the same as the group of translational symmetries:

$$\Gamma^+(F_3) = \Delta(F_3) = T_1.$$

The base rectangle for this frieze is identical to the rectangular pattern R_3 which we studied in Exercise 1.6. There we saw that its group of symmetries is Q_h, which is generated by the reflection h in its horizontal axis of symmetry, as Figure 2.10 illustrates.

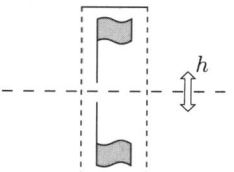

Figure 2.10

If we now consider the entire frieze, we see that h is indeed a symmetry of it, as Figure 2.11 illustrates.

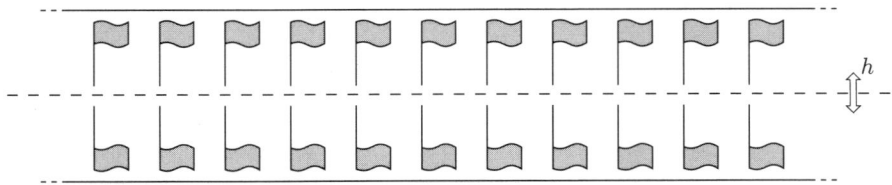

Figure 2.11

Since the composite of two symmetries is again a symmetry, we can take this reflection h along with all the translations and form the group

$$\{xy : \; x \in T_1, \; y \in Q_h, \; ht = th\}.$$

This set can be shown to be a group in a similar fashion to the corresponding set in the case of F_2.

The relation $ht = th$ was used in our description of E_1 in Subsection 1.1. You can check this relation using the Frieze Card and the Overlay for Side 1. Alternatively, if you choose a coordinate system with x-axis the centre line, you can derive the relation from Equation 6b of the Isometry Toolkit by noting that $t = t[\mathbf{a}]$, $h = q[0]$ and $q[0](\mathbf{a}) = \mathbf{a}$.

You checked the same relation in Unit IB2, Exercise 1.10(b).

We know that this is a subgroup of the group of symmetries of F_3. That is,

$$\Gamma(F_3) \supseteq \{xy : \; x \in T_1, \; y \in Q_h; \; ht = th\}.$$

It remains to determine whether there are any further symmetries of F_3.

For each integer n, we temporarily denote by g_n the group element $t^n h$. Thus g_0 is just the reflection h in the centre line, while, for $n \neq 0$, g_n is the glide reflection consisting of reflection in the centre line followed by translation by n steps to the right, as Figure 2.12 illustrates.

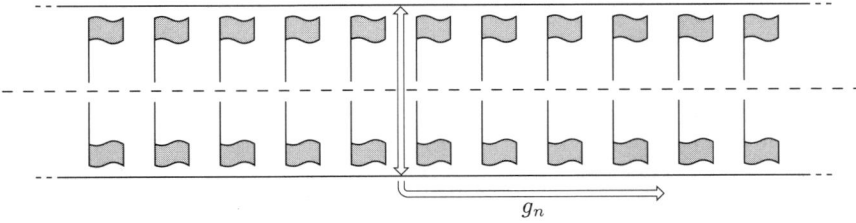

Figure 2.12

We have thus accounted for all translational symmetries of F_3, for the reflection symmetry in the centre line, and for all the glide reflections whose axis is the centre line. Now all the flags have their masts on the left, and under rotations, or reflections in vertical axes, they would have to map to flags with masts on the right. Therefore there are no such symmetries of F_3, and so by Theorem 1.2 we have accounted for all the symmetries of F_3. Thus we now know that

$$\Gamma(F_3) = \{xy : \; x \in T_1, \; y \in Q_h; \; ht = th\}.$$

The expression xy is the **standard form** of an element of $\Gamma(F_3)$, and our definition above shows us how to express in standard form the reflection and glide reflections which we temporarily denoted by g_0 and g_n.

Exercise 2.8

Express in standard form the result of performing the translation t^n followed by the glide reflection g_i.

Exercise 2.9

Express in standard form the result of performing the glide reflection g_i followed by the glide reflection g_j.

Exercise 2.10

Express in explicit form the glide reflection g_n.

2.3 Type 4

Consider the symmetries of our standard Type 4 frieze, F_4, shown in Figure 2.13.

This is the last frieze on Side 1 of the Frieze Card and on the corresponding overlay.

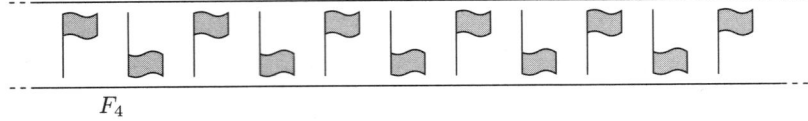

F_4

Figure 2.13

As usual, the group of translational symmetries is

$$\Delta(F_4) = T_1,$$

where the generating translation t moves each flag two places to the right, as shown in Figure 2.14.

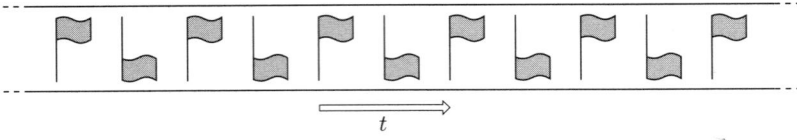

Figure 2.14

There are no rotational symmetries, and so the group of direct symmetries is the same as the group of translational symmetries:

$$\Gamma^+(F_4) = \Delta(F_4) = T_1.$$

As with friezes of Type 2 and Type 3, we can consider the translations as moving blocks of two flags, so that the base rectangle for this frieze (shown in Figure 2.15) is identical to the rectangular pattern R_4 which we studied in Exercise 1.6.

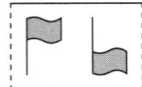

Figure 2.15

An alternative choice of base rectangle is shown below. Again, either choice would result in the same description of $\Gamma(F_4)$.

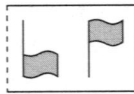

In Exercise 1.6 we saw that the group of symmetries of R_4 is trivial, containing just the identity e. Therefore, in this case, the interplay between the group of translations and the symmetries of the base rectangle gives us nothing new.

However, when we consider geometrically all the possible symmetries of this frieze, we observe that, as well as the translations, we also have glide reflections along the centre line, as illustrated in Figure 2.16.

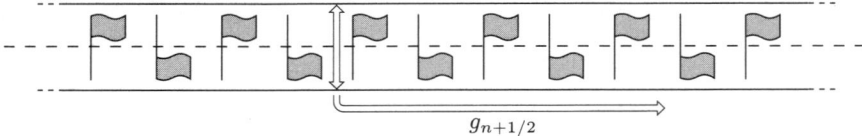

Figure 2.16

Each such glide reflection is the reflection h in the centre line, followed by a translation through an integer number of steps and then a further half step to the right. In the case of n steps (plus half a step to the right), we shall write this glide reflection as $g_{n+1/2}$. For example, the glide reflection $g_{1/2}$, shown in Figure 2.17, shifts only half a step to the right, and the glide reflection $g_{-1/2}$ shifts only half a step to the left.

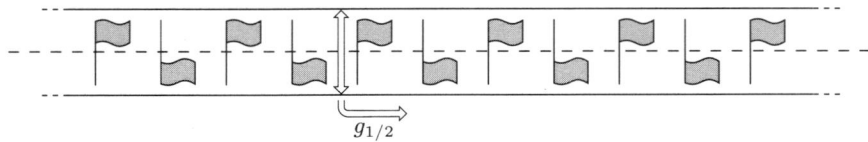

Figure 2.17

We now need notation for translation half a step to the right. We shall denote it by $t_{1/2}$, where

$$t_{1/2}\, t_{1/2} = t.$$

The glide reflections are now described as follows:

$$g_{n+1/2} = t^n\, t_{1/2}\, h.$$

The problem with this description of the symmetries is that it is in terms of the translational symmetries t^n and $t_{1/2}$ and the reflection h, where neither of the last two are symmetries of the frieze. However, if we use $g_{1/2}$ for the simplest glide reflection, we can write the general glide reflection $g_{n+1/2}$ in the **standard form**

$$t^n g_{1/2}, \qquad \text{where } g_{1/2} = t_{1/2}h.$$

Now that we know how to represent the symmetries algebraically, we need to be able to use the algebra to evaluate composites. We already know the behaviour of the translations. We only need to know the relations between the glide reflections and the translations in order to obtain composites in standard form.

Example 2.3

To express the composite of t followed by $g_{1/2}$ in standard form, we set

$$\begin{aligned}
g_{1/2}\, t &= t_{1/2}\, h\, t && \text{(since } g_{1/2} = t_{1/2}\, h\text{)} \\
&= t_{1/2}\, t\, h && \text{(since } ht = th\text{)} \\
&= t\, t_{1/2}\, h && \text{(since, with } t = t[\mathbf{a}]\text{, both } t_{1/2}\, t \text{ and } t\, t_{1/2} \text{ equal the translation by } \tfrac{3}{2}\mathbf{a}\text{)} \\
&= t\, g_{1/2}.
\end{aligned}$$

♦

Exercise 2.11

Express the composite of $g_{1/2}$ with itself in standard form.

We have now accounted for all translational and glide reflection symmetries of F_4, and must check that these indeed account for the whole of $\Gamma(F_4)$. As with F_3, the other possibilities (according to Theorem 1.2) are rotations by π and reflections in vertical axes. Just as with F_3, however, all the flags of F_4 have their masts on the left, and so there are no such symmetries. Thus, we now know that

$$\Gamma(F_4) = \left\{ x,\ x g_{1/2} : \ x \in T_1;\ g_{1/2}^2 = t,\ g_{1/2} t = t g_{1/2} \right\}.$$

We have noted that h, the reflection in the centre line (i.e. the 'trivial glide reflection') is not a symmetry of F_4.

Here we have something slightly different from before. Instead of having the group of symmetries described in terms of two simpler groups, we now have a description in terms of the group of translations T_1 and a single element $g_{1/2}$ whose square is a translation. The expressions x and $x g_{1/2}$ are the two possible **standard forms** of elements of $\Gamma(F_4)$.

Exercise 2.12

Use the algebraic representation of the group of symmetries to show that

$$t^{-1} g_{1/2}^2 = e$$

and hence that

$$g_{1/2}^{-1} = t^{-1} g_{1/2}.$$

Use the algebra to show further that

$$g_{n+1/2}^{-1} = t^{-(n+1)} g_{1/2}.$$

Interpreting the results in Exercise 2.12 geometrically, we have

$$g_{1/2}^{-1} = t^{-1} g_{1/2} = g_{-1+(1/2)} = g_{-1/2}$$

and

$$g_{n+1/2}^{-1} = t^{-(n+1)} g_{1/2} = g_{-(n+1)+(1/2)} = g_{-n-1/2}.$$

Thus the inverse of a glide reflection is a glide reflection with the glide distance reversed, as one would intuitively expect.

3 FRIEZES WITH ROTATIONAL SYMMETRY

So far we have dealt with friezes which exhibit only translational and indirect symmetries. Now we shall tackle those which exhibit rotational symmetries. The more complicated of these will have both rotational symmetries and indirect symmetries.

3.1 Type 5

Geometry

Consider the symmetries of our standard Type 5 frieze, F_5, shown in Figure 3.1.

This is the first frieze on Side 2 of the Frieze Card and on the corresponding overlay.

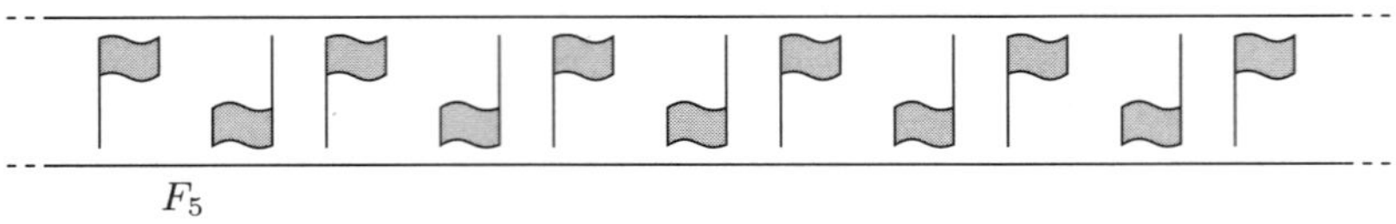

Figure 3.1

As always, the group of translational symmetries is

$$\Delta(F_5) = T_1,$$

where the generating translation t moves each flag two places to the right, as shown in Figure 3.2.

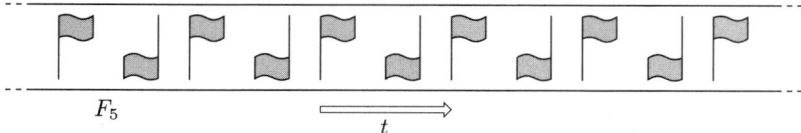

Figure 3.2

However, there are now rotational symmetries, since if you rotate the frieze through a half-turn about any point P midway between two flags, as in Figure 3.3 for example, you will observe that the result is the same frieze.

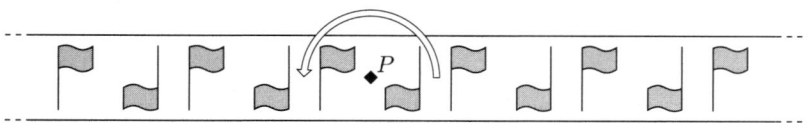

Figure 3.3

We conventionally use a small diamond to indicate a centre of rotation of order 2, where a rotation has *order n* if n is the smallest positive integer such that repeating the rotation n times gives the identity isometry.

Hence the group of direct symmetries must be larger than the group of translational symmetries:

$$\Gamma^+(F_5) \supset \Delta(F_5) = T_1,$$

where the $\supset$ symbol means that $\Gamma^+(F_5)$ is strictly larger than $\Delta(F_5)$, i.e. that $\Delta(F_5)$ is a strict subset of $\Gamma^+(F_5)$, so that the two groups cannot be equal.

As with the friezes in the last section, we can consider the translations as moving blocks of two flags, so that the base rectangle for the frieze is identical to the rectangular pattern R_5 which we studied in Exercise 1.6. There we saw that its group of symmetries is R_r, which is generated by a rotation r through π about the centre of the rectangle, as illustrated in Figure 3.4.

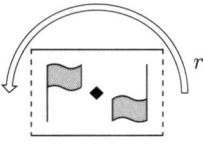

Figure 3.4

An alternative choice of base rectangle is shown below. Again, either choice would result in the same description for $\Gamma(F_5)$.

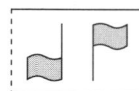

If we now consider the entire frieze, we see that r is indeed a symmetry of it (see Figure 3.3).

Since the composite of two symmetries is again a symmetry, we can take this rotation r, along with all the translations, and form the group of symmetries

$$\{xy: \; x \in T_1, \; y \in R_r; \; rt = t^{-1}r\}.$$

You can check that the relation $rt = t^{-1}r$ holds using the Frieze Card and the Overlay for Side 2. You can also derive this relation from Equation 6a of the Isometry Toolkit, by noting that $t = t[\mathbf{a}]$, $r = r[\pi]$ and $r[\pi](\mathbf{a}) = -\mathbf{a}$.

This set can be shown to be a group in a similar fashion to the corresponding set in the case of F_2.

You checked the same relation in *Unit IB2*, Exercise 1.10(a).

We know that this is a subgroup of the group of symmetries of F_5. That is,

$$\Gamma(F_5) \supseteq \{xy: \; x \in T_1, \; y \in R_r; \; rt = t^{-1}r\}.$$

As before, it remains to determine whether there are any further symmetries of F_5.

Exercise 3.1

Write down, in terms of t and r, all the symmetries in the above subgroup.

We have considered the rotational symmetry r which rotates through π about the centre of our base rectangle. But any translate of this rectangle might easily have been chosen as the base rectangle, and so clearly there is a rotational symmetry about the centre of each of these translates. We shall temporarily write r_n for the symmetry derived from the rectangle translated n steps to the right ($n \in \mathbb{Z}$). Explicitly this gives symmetries

$$\ldots, r_{-2}, r_{-1}, r_0, r_1, r_2, \ldots$$

where r_0 is the basic rotation r.

Further inspection of the frieze shows that there are also rotational symmetries about the midpoints of the vertical axes separating translates of our original base rectangle. These points are found by translating the centre of the base rectangle through an integer number of steps and then a further half step to the right. We shall temporarily write $r_{n+1/2}$ for the rotational symmetry derived by translating the base rectangle n steps and then a further half step to the right. Explicitly this gives symmetries

$$\ldots, r_{-3/2}, r_{-1/2}, r_{1/2}, r_{3/2}, r_{5/2}, \ldots$$

where, for example, $r_{1/2}$ and $r_{-1/2}$ are rotations about the midpoints of the lines forming the vertical sides of the base rectangle.

The rotation centres for several of these rotations are marked in Figure 3.5.

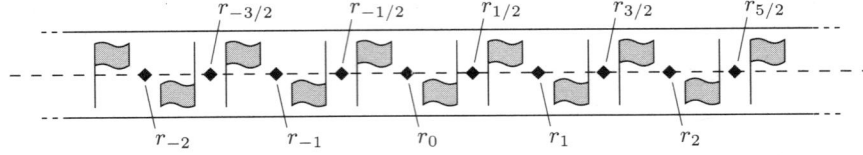

Figure 3.5

Exercise 3.2

What is the composite symmetry $r_{-1/2} r_{1/2}$?

You may find it useful to use the Frieze Card and the Overlay for Side 2. Alternatively, you may prefer to use the Isometry Toolkit.

Algebra

As before, we set up a coordinate system with its origin at the centre of the base rectangle and the x-axis pointing horizontally to the right along the frieze. With $t = t[\mathbf{a}]$, we can argue as in Subsection 2.1 to obtain

$$r_n = r[n\mathbf{a}, \pi] \tag{3.1}$$

and

$$r_{n+1/2} = r\left[\left(n + \tfrac{1}{2}\right)\mathbf{a}, \pi\right]. \tag{3.2}$$

Equation 9 of the Isometry Toolkit and the fact that $r = r[\pi]$ immediately allow us to deduce that:

$$\begin{aligned} r_n &= t[2n\mathbf{a}]\, r[\pi] \\ &= t^{2n} r; \end{aligned} \tag{3.3}$$

$$\begin{aligned} r_{n+1/2} &= t[(2n+1)\mathbf{a}]\, r[\pi] \\ &= t^{2n+1} r. \end{aligned} \tag{3.4}$$

We shall now consider the explicit forms of these symmetries, assuming (as before) that $\mathbf{a} = (a, 0)$.

We have already seen (in Exercise 2.3) that

$$t^n : (x, y) \mapsto (x + na, y).$$

Exercise 3.3

Write out the explicit forms of r_n and $r_{n+1/2}$.

Next, as in Subsection 2.1, we ask you to show that the symmetries t^n, r_n and $r_{n+1/2}$, for all $n \in \mathbb{Z}$, constitute the group we defined earlier. We also ask you to show that this group is indeed $\Gamma(F_5)$.

Exercise 3.4

Show that the set $\{t^n, r_n, r_{n+1/2} : n \in \mathbb{Z}\}$ is exactly the group

$$\{xy : x \in T_1,\ y \in R_r;\ rt = t^{-1}r\}.$$

Exercise 3.5

By observing the shape of the flags, conclude that the above group is the whole symmetry group of F_5.

Thus we now know that

$$\Gamma(F_5) = \{xy : x \in T_1,\ y \in R_r;\ rt = t^{-1}r\}.$$

As before, the expression xy (where $x \in T_1$ and $y = e$ or r) for an element of $\Gamma(F_5)$ is the **standard form** of that element.

Exercise 3.6

Express in standard form the result of performing the translation t^n followed by the rotation r_i.

Exercise 3.7

Express in standard form the result of performing the rotation r_i followed by the rotation r_j.

The whole of the argument (including the exercises) concerning $\Gamma(F_5)$ bears a remarkable similarity to that concerning $\Gamma(F_2)$! This is not surprising, since the *algebraic* behaviour of these two groups is exactly the same; the role played by the rotation r in $\Gamma(F_5)$ is identical to the role played by the reflection v in $\Gamma(F_2)$. We say that two such groups are *isomorphic*, and we write

$$\Gamma(F_5) \cong \Gamma(F_2).$$

Of course, we can still tell these groups apart *geometrically*, since one contains rotations but no reflections while the other contains reflections but no rotations.

Isomorphisms will be dealt with more formally in *Unit IB4*.

3.2 Type 6

Consider the symmetries of our standard Type 6 frieze, F_6, shown in Figure 3.6.

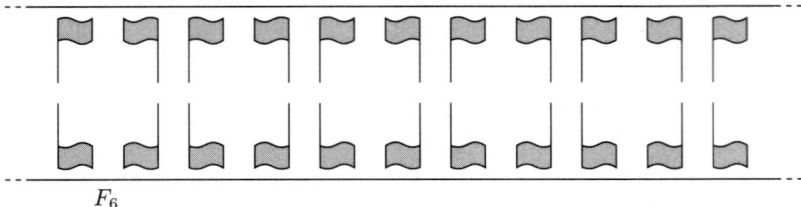

F_6

Figure 3.6

At the risk of becoming monotonous, we still know that the group of translational symmetries is

$$\Delta(F_6) = T_1,$$

where the generating translation t moves each column of flags two places to the right, as shown in Figure 3.7.

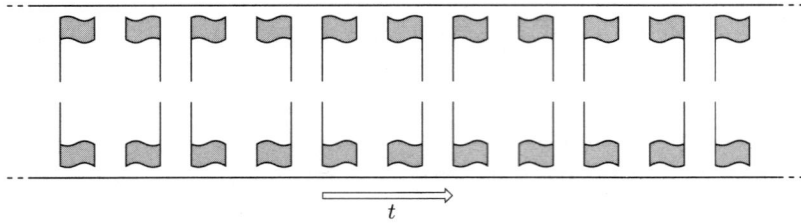

Figure 3.7

There are two types of reflection symmetries, reflecting in the centre line and in the vertical lines halfway between columns of flags, as well as rotational symmetries, centred on points midway between columns of flags, as Figure 3.8 illustrates.

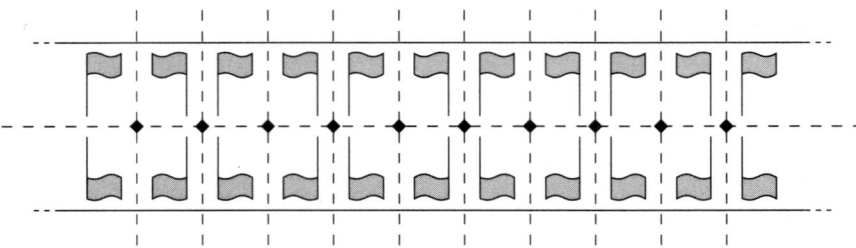

Figure 3.8

Hence the best we can currently say is that there are the strict inclusions

$$\Gamma(F_6) \supset \Gamma^+(F_6) \supset \Delta(F_6) = T_1.$$

The base rectangle for this frieze is identical to the rectangular pattern R_6 which we studied in Exercise 1.6. There we saw that its group of symmetries is $\Gamma(\square) = V$, the entire group of symmetries of a rectangle, as Figure 3.9 illustrates.

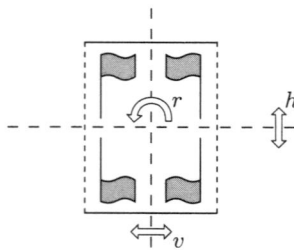

Figure 3.9

An alternative choice of base rectangle is shown below. Again, either choice would result in the same description for $\Gamma(F_6)$.

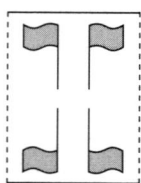

The symmetries in V are symmetries of the entire frieze, and along with all the translations they form the group of symmetries

$$\{xy:\ x \in T_1,\ y \in V;\ ht = th,\ vt = t^{-1}v\}.$$

From previous sections, we are able to recognize geometrically what each of the symmetries in this group is. For $n \in \mathbb{Z}$:

- t^n — a translation to the right through n steps;
- $t^{2n}r$ — a rotation about a centre n steps to the right (written r_n);
- $t^{2n+1}r$ — a rotation about a centre midway between the rectangles n and $n+1$ steps to the right (written $r_{n+1/2}$);
- $t^{2n}v$ — a reflection in the vertical axis of symmetry of the rectangle n steps to the right (written v_n);
- $t^{2n+1}v$ — a reflection in the line midway between the rectangles n and $n+1$ steps to the right (written $v_{n+1/2}$);
- $t^0 h = h$ — a reflection in the centre line (written g_0);
- $t^n h$ ($n \neq 0$) — a glide reflection consisting of reflection in the centre line followed by translation through n steps to the right (written g_n).

Note that this group is the same as the group E_1 of symmetries of a plain rectangular frieze.

Note that $t^0 = e$.

All the types of frieze symmetries covered by Theorem 1.2 are present here, and it is geometrically clear that within each type we have found all possible symmetries. Thus we now know that

$$\Gamma(F_6) = \{xy:\ x \in T_1,\ y \in V;\ ht = th,\ vt = t^{-1}v\} = E_1.$$

Once again, the expression xy is the **standard form** of an element of $\Gamma(F_6)$.

Exercise 3.8

Describe $\Gamma^+(F_6)$.

Exercise 3.9

Express in standard form the result of performing the glide reflection g_i followed by the rotation r_j.

3.3 Type 7

Geometry

Consider the symmetries of our standard Type 7 frieze, F_7, shown in Figure 3.10.

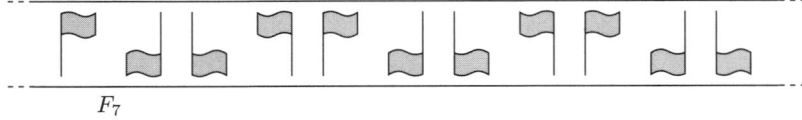

Figure 3.10

This will turn out to be a rather peculiar case.

As always, the group of translational symmetries is

$$\Delta(F_7) = T_1,$$

where the generating translation t moves each flag four places to the right, as shown in Figure 3.11.

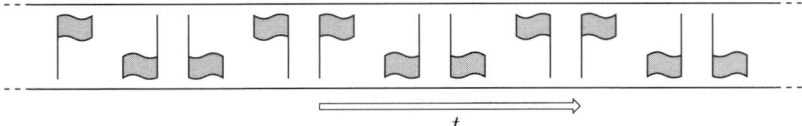

Figure 3.11

As with Type 6, there are both rotational and reflection symmetries, and so we have the strict inclusions

$$\Gamma(F_7) \supset \Gamma^+(F_7) \supset \Delta(F_7) = T_1.$$

The base rectangle for this frieze is identical to the rectangular pattern R_7 which we studied in Exercise 1.6. There we saw that its group of symmetries is Q_v, which is generated by a reflection v in the vertical axis of symmetry of the rectangle, as illustrated in Figure 3.12.

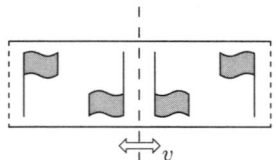

Figure 3.12

An alternative choice of base rectangle, that would result in the same description of $\Gamma(F_7)$, is shown below.

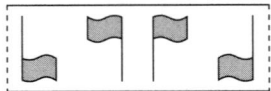

Again, if we now consider the entire frieze, we see that v is indeed a symmetry of it, as illustrated in Figure 3.13.

Note that, as we shall see shortly, there are two other choices of base rectangle, which would result in different descriptions of $\Gamma(F_7)$.

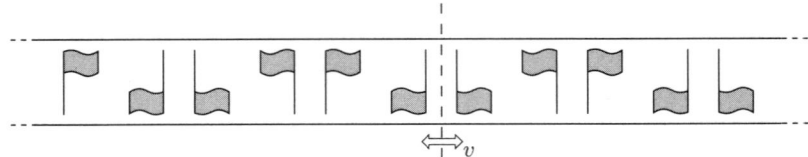

Figure 3.13

Since the composite of two symmetries is again a symmetry, we can take this reflection v, along with all the translations, and form the group of symmetries

$$\{xy: \; x \in T_1, \; y \in Q_v; \; vt = t^{-1}v\}.$$

We certainly know that this is a subgroup of the group of symmetries of F_7. That is,

$$\Gamma(F_7) \supseteq \{xy: \; x \in T_1, \; y \in Q_v; \; vt = t^{-1}v\}.$$

We noted that this is a group and that the relation $vt = t^{-1}v$ holds in Subsection 2.1.

However, although the rectangular pattern R_7 does not exhibit a rotational symmetry, it has two *subsets* (of two flags each) that do exhibit rotational symmetry, as Figure 3.14 illustrates.

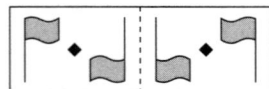

Figure 3.14

Furthermore, though these are *not* symmetries of the base rectangle of F_7, they *are* symmetries of the entire frieze, as Figure 3.15 illustrates.

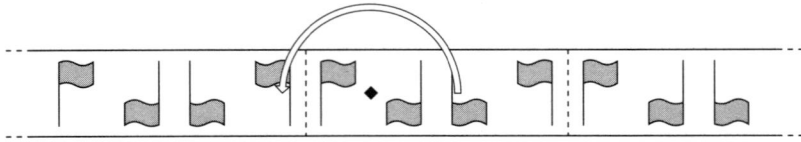

Figure 3.15

This is similar to the way glide reflections appeared for our Type 4 frieze, where they did not appear as symmetries of the base rectangle. In fact, this Type 7 frieze also has glide reflections which do not come from symmetries of the base rectangle, as Figure 3.16 illustrates.

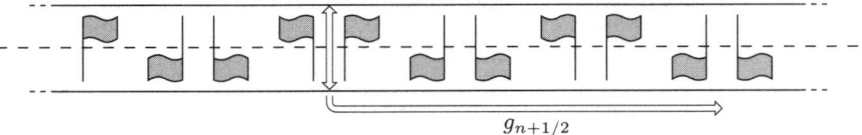

Figure 3.16

The way the translations, rotations, reflections and glide reflections interact will result in a more complicated representation of the group of symmetries of the frieze than those we have met so far. In order to arrive at this representation, we next consider the geometry of the various types of symmetry.

- *Translations*   As usual, we have the translations t^n ($n \in \mathbb{Z}$) taking us n steps to the right, where $t^0 = e$.
- *Reflections in vertical axes*   We have the two families of such symmetries:

$$\ldots, v_{-2}, v_{-1}, v_0, v_1, v_2, \ldots$$

and

$$\ldots, v_{-3/2}, v_{-1/2}, v_{1/2}, v_{3/2}, v_{5/2}, \ldots$$

These reflect in translates of the vertical axis of symmetry of the base rectangle and in the lines between rectangles. We met these when we studied our Type 2 and Type 6 friezes. The former can be written

$$v_n = t^{2n}v \quad (n \in \mathbb{Z}), \qquad \text{where } v_0 = v,$$

and the latter

$$v_{n+1/2} = t^{2n+1}v \quad (n \in \mathbb{Z}).$$

- *Glide reflections*   The first fact to note is that the reflection h in the centre line is not itself a symmetry of the frieze. However, the glide reflections which first reflect and then translate through an integer number of steps and then a further half step to the right *are* symmetries. We met these when we studied our Type 4 frieze. We saw that we could write these as

$$g_{n+1/2} = t^n g_{1/2} \quad (n \in \mathbb{Z}),$$

where the basic glide reflection $g_{1/2}$ moves half a step to the right.

- *Rotations*   If we look at the base rectangle, we see that there is a rotation about a point one quarter of a step to the right of the centre of that rectangle which is a symmetry of the entire frieze. Similarly there is a rotational symmetry about a point one quarter of a step to the left of the centre of the base rectangle. If we look at translates of the base rectangle, we find similar centres of rotation. We thus obtain a family of rotations, which can be written

$$\ldots, r_{-3/4}, r_{-1/4}, r_{1/4}, r_{3/4}, r_{5/4}, \ldots$$

The rotation centres for some of these are shown in Figure 3.17.

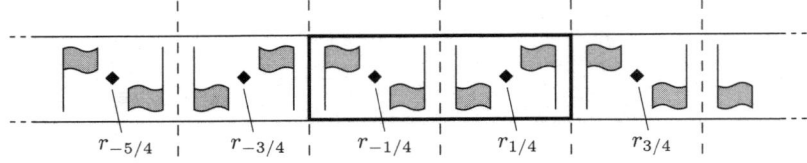

Figure 3.17

The general rotation has the form

$$r_{n/2+1/4} \quad (n \in \mathbb{Z}),$$

that is, its index is half an integer plus a quarter. The most obvious example is the rotation with index $\frac{1}{4}$, obtained by putting $n = 0$, namely $r_{1/4}$. We can compose this symmetry, $r_{1/4}$, with translations to get all of the other rotations. We have seen, when studying our Type 5 frieze, that the effect of composing a rotation with a translation along the frieze is to give another rotation with centre moved half the distance of the translation. Hence, for the general rotational symmetry of this frieze,

$$r_{n/2+1/4} = t^n r_{1/4} \quad (n \in \mathbb{Z}).$$

In the above geometric description, it has been harder to deal with the rotational symmetries than the reflection symmetries. This is not an intrinsic property of the frieze but a consequence of our choice of base rectangle. If, for example, we had chosen instead one of the base rectangles in Figure 3.18(a), then the rotational symmetries would have occurred at the centres of the rectangles and between them, as Figure 3.18(b) illustrates; and (as for Type 5 friezes) would have been denoted by r_n and $r_{n+1/2}$. The vertical reflection axes, however, would have occurred one quarter and three quarters of the way along each rectangle (see Figure 3.18(b)), so the overall complexity would have been the same as for the representation we have chosen.

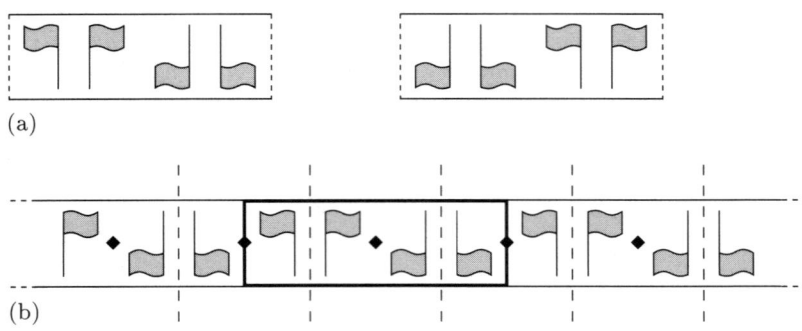

(a)

(b)

Figure 3.18

Algebra

We have now exhausted all possible geometric symmetries of the frieze F_7, and so we have the set

$$\{t^n, \; t^n v, \; t^n g_{1/2}, \; t^n r_{1/4}\}.$$

If we try to represent the group of symmetries of F_7 this way, we obtain

$$\Gamma(F_7) = \{t^n, \; t^n v, \; t^n g_{1/2}, \; t^n r_{1/4} : \; \ldots \; ? \; \ldots\}$$

or, more succinctly, using all the relations that we already know:

$$\Gamma(F_7) = \{xy, \; xg_{1/2}, \; xr_{1/4} : x \in T_1, \; y \in Q_v;$$
$$vt = t^{-1}v,$$
$$g_{1/2}^2 = t, \; g_{1/2} t = t g_{1/2}, \; \ldots \; ? \; \ldots \}.$$

This still leaves us with the problem of how the rotation $r_{1/4}$ interacts with the translation t, the reflection v, the glide reflection $g_{1/2}$ and itself.

Let us write t, v, $g_{1/2}$ and $r_{1/4}$ in the standard form suitable for manipulation by the Isometry Toolkit. As usual, we place the origin at the centre of the base rectangle, the x-axis along the centre line and the y-axis along the vertical axis of symmetry of the base rectangle. We use the vector $\mathbf{a} = (a, 0)$ to denote translation by one step, so that

$$t = t[\mathbf{a}]. \tag{3.5}$$

Since $v = v_0$, we know from Equation 2.1 of Subsection 2.1 that

$$v = q[\pi/2]. \tag{3.6}$$

Futhermore, $g_{1/2}$ is obtained by reflecting in the x-axis followed by translating by $\tfrac{1}{2}a$ to the right, so

$$g_{1/2} = t[\tfrac{1}{2}\mathbf{a}] \, q[0]. \tag{3.7}$$

Now how about $r_{1/4}$? This is rotation through π about $\tfrac{1}{4}\mathbf{a}$, and so, using Equation 9 of the Isometry Toolkit,

$$\begin{aligned} r_{1/4} &= r[\tfrac{1}{4}\mathbf{a}, \pi] \\ &= t[\tfrac{1}{2}\mathbf{a}] \, r[\pi]. \end{aligned} \tag{3.8}$$

Example 3.1

We can describe the symmetries $g_{1/2} \, v$ and $v \, g_{1/2}$ in terms of t and $r_{1/4}$ as follows:

$$\begin{aligned} g_{1/2} \, v &= t[\tfrac{1}{2}\mathbf{a}] \, q[0] \, q[\pi/2] && \text{(by Equations 3.6 and 3.7)} \\ &= t[\tfrac{1}{2}\mathbf{a}] \, r[-\pi] && \text{(by Equation 3 of the Isometry Toolkit)} \\ &= t[\tfrac{1}{2}\mathbf{a}] \, r[\pi] \\ &= r_{1/4} && \text{(by Equation 3.8);} \end{aligned}$$

$$\begin{aligned} v \, g_{1/2} &= q[\pi/2] \, t[\tfrac{1}{2}\mathbf{a}] \, q[0] && \text{(by Equations 3.6 and 3.7)} \\ &= t[-\tfrac{1}{2}\mathbf{a}] \, q[\pi/2] \, q[0] && \text{(by Equation 6b of the Toolkit)} \\ &= t[-\tfrac{1}{2}\mathbf{a}] \, r[\pi] && \text{(by Equation 3 of the Toolkit)} \\ &= t[-\mathbf{a}] \, t[\tfrac{1}{2}\mathbf{a}] \, r[\pi] \\ &= t^{-1} r_{1/4} && \text{(by Equations 3.5 and 3.8).} \end{aligned}$$ ♦

Exercise 3.10 ─────────────────────────────────

Re-derive the relations $g_{1/2} \, v = r_{1/4}$ and $v \, g_{1/2} = t^{-1} r_{1/4}$ by putting $t, v, g_{1/2}$ and $r_{1/4}$ into explicit form.

───

The first of these two relations,

$$g_{1/2} \, v = r_{1/4},$$

enables us to remove the symbol $r_{1/4}$ from the representation of the group. We have

$$\Gamma(F_7) = \{xy, \, x \, g_{1/2}, \, x \, g_{1/2} \, v : \, x \in T_1, \, y \in Q_v; \, \ldots \, ? \, \ldots\}$$

and thence

$$\Gamma(F_7) = \{xy, \, x \, g_{1/2} \, y : \, x \in T_1, \, y \in Q_v; \, \ldots \, ? \, \ldots\},$$

where y can take the value e or v.

The second of the two relations,

$$v \, g_{1/2} = t^{-1} r_{1/4},$$

combines with the first to give the relation

$$v \, g_{1/2} = t^{-1} g_{1/2} \, v,$$

and this provides us with just enough information to rearrange combinations of v and $g_{1/2}$ into standard form.

Hence the representation of the group of symmetries which we are after is

$$\Gamma(F_7) = \{xy, \, x\,g_{1/2}\,y : \, x \in T_1, \, y \in Q_v;$$
$$vt = t^{-1}v,$$
$$g_{1/2}^2 = t, \, g_{1/2}\,t = t\,g_{1/2},$$
$$v\,g_{1/2} = t^{-1}\,g_{1/2}\,v\}.$$

We have written this representation on several lines to show how the relations should be grouped.

The expressions xy and $x\,g_{1/2}\,y$ are the two possible **standard forms** of elements of $\Gamma(F_7)$.

Exercise 3.11

Express in standard form the result of performing the glide reflection $g_{i+1/2}$ followed by the rotation $r_{j/2+1/4}$.

4 CLASSIFYING FRIEZE PATTERNS

4.1 Isomorphism properties of frieze groups

In the previous sections we have found the groups of symmetries of the seven types of frieze:

$\Gamma(F_1) = T_1$

$\Gamma(F_2) = \{xy : \, x \in T_1, \, y \in Q_v; \, vt = t^{-1}v\}$

$\Gamma(F_3) = \{xy : \, x \in T_1, \, y \in Q_h; \, ht = th\}$

$\Gamma(F_4) = \{x, \, x\,g_{1/2} : \, x \in T_1; \, g_{1/2}^2 = t, \, g_{1/2}\,t = t\,g_{1/2}\}$

$\Gamma(F_5) = \{xy : \, x \in T_1, \, y \in R_r; \, rt = t^{-1}r\}$

$\Gamma(F_6) = \{xy : \, x \in T_1, \, y \in V; \, ht = th, \, vt = t^{-1}v\}$

$\Gamma(F_7) = \{xy, \, x\,g_{1/2}\,y : \, x \in T_1, \, y \in Q_v; \, vt = t^{-1}v, \, g_{1/2}^2 = t, \, g_{1/2}\,t = t\,g_{1/2}, \, v\,g_{1/2} = t^{-1}\,g_{1/2}\,v\}$

We shall now make some general observations about the algebraic properties of these groups.

First of all, remember our discussion at the end of Subsection 3.1, where we noted that the groups $\Gamma(F_2)$ and $\Gamma(F_5)$ are isomorphic, i.e. they are *algebraically* identical. We demonstrated this by noting that the role played by v in $\Gamma(F_2)$ is algebraically identical to the role played by r in $\Gamma(F_5)$, so that if we identify v with r then the groups are the same. However, we also noted that the groups are *geometrically* distinct. We can see this by looking at their subgroups of direct symmetries. We find that

$$\Gamma^+(F_2) = T_1 \quad \text{and} \quad \Gamma^+(F_5) = F_5,$$

and these are clearly distinct.

Exercise 4.1

Which other frieze group is isomorphic to $\Gamma(F_1) = T_1$, i.e. which other frieze group has a single element all of whose powers are distinct and which generates the group?

Though $\Gamma(F_1)$ and $\Gamma(F_4)$ are isomorphic, i.e. they are algebraically identical, they are geometrically distinct, since one contains glide reflections and the other doesn't.

Are any of the other frieze groups isomorphic to each other? The best way to approach this is first of all to identify those groups which cannot be isomorphic. Now, one way of showing that two groups are *not* isomorphic is to find an algebraic property which one has but the other doesn't. For example, some groups are Abelian and some are not.

Exercise 4.2

Which of the frieze groups are Abelian?

Remember, from *Unit IB2*, that a group is *Abelian* if $xy = yx$ for all pairs x, y of elements in the group.

As a result of Exercise 4.2, we know that none of $\Gamma(F_1), \Gamma(F_3)$ or $\Gamma(F_4)$ is isomorphic to any of $\Gamma(F_2)$, $\Gamma(F_5)$, $\Gamma(F_6)$ or $\Gamma(F_7)$.

Another important property of groups is the existence of elements of particular orders. If one group has an element of a particular order and the other doesn't, then they cannot be isomorphic.

Remember from *Unit IB2* that an element x has *order* n if n is the smallest positive integer such that $x^n = e$, and that x has *infinite order* if no such n exists.

Exercise 4.3

Show that $\Gamma(F_1)$ and $\Gamma(F_3)$, both of which are Abelian, are not isomorphic.

We shall now make rather a surprising observation. Despite the fact that $\Gamma(F_7)$ has the most complicated *geometric* description of all the frieze groups, it is in fact isomorphic to $\Gamma(F_2)$ and $\Gamma(F_5)$!

To see this, look again at the relations which define $\Gamma(F_7)$. They include the relation

$$g_{1/2}^2 = t.$$

Thus t can be replaced by $g_{1/2}^2$ in all the other relations, to obtain:

$$v g_{1/2}^2 = g_{1/2}^{-2} v;$$
$$g_{1/2} g_{1/2}^2 = g_{1/2}^2 g_{1/2};$$
$$v g_{1/2} = g_{1/2}^{-2} g_{1/2} v$$
$$= g_{1/2}^{-1} v.$$

Since $g_{1/2} g_{1/2}^2 = g_{1/2}^2 g_{1/2}$ is automatically true, while $v g_{1/2}^2 = g_{1/2}^{-2} v$ can be deduced from $v g_{1/2} = g_{1/2}^{-1} v$, it follows that this last relation is all we need. Thus we can write

$$\Gamma(F_7) = \left\{ xy, \; x g_{1/2} y : \; x \in T_1, \; y \in Q_v; \; v g_{1/2} = g_{1/2}^{-1} v \right\}.$$

But since x is just a power of t (and hence an even power of $g_{1/2}$), we can write this in turn as

$$\Gamma(F_7) = \left\{ g_{1/2}^{2m} y, \; g_{1/2}^{2m+1} y : \; m \in \mathbb{Z}, \; y \in Q_v; \; v g_{1/2} = g_{1/2}^{-1} v \right\}$$
$$= \left\{ g_{1/2}^n y : \; n \in \mathbb{Z}, \; y \in Q_v; \; v g_{1/2} = g_{1/2}^{-1} v \right\},$$

and this is clearly of the same algebraic form as $\Gamma(F_2)$, with $g_{1/2}$ replacing t.

What about $\Gamma(F_6)$? Is this also isomorphic to $\Gamma(F_2)$, $\Gamma(F_5)$ and $\Gamma(F_7)$? The answer is *no*, because $\Gamma(F_6)$ contains three elements of order 2, namely r, h and v, such that the product of any two is equal to the third. However, if we look (for example) at $\Gamma(F_2)$, the only elements of order 2 are vertical reflections (of the form $t^n v$), and when we combine two of these we obtain a translation, which is of infinite order.

Thus our final classification of the isomorphism properties of the frieze groups is as follows.

Theorem 4.1 Isomorphism properties of frieze groups

The frieze groups $\Gamma(F_1), \ldots, \Gamma(F_7)$ belong to the following four isomorphism classes:

Abelian

$\Gamma(F_1) \cong \Gamma(F_4)$
$\Gamma(F_3)$

Non-Abelian

$\Gamma(F_2) \cong \Gamma(F_5) \cong \Gamma(F_7)$
$\Gamma(F_6)$

In other words, although there are *seven* geometrically distinct frieze groups, there are only *four* algebraically distinct such groups.

4.2 The frieze group algorithm

In this subsection we shall see how the properties of the symmetry groups of friezes lead to the design of the algorithm that enables us to determine the type of any given frieze.

In the previous sections we have seen examples of seven types of frieze. We have found their groups of symmetries, and what we mean by the type of a frieze pattern is precisely that — its symmetry group. We have to be careful, since we have seen that there are a pair and a triple of groups that are isomorphic as abstract groups, but when we look at the geometric nature of the elements we can see that they are different. Hence an algorithm which distinguishes types of frieze must be developed from our knowledge of the occurrence of different *geometric* types of symmetry in its group of symmetries.

This algorithm is also described in the video programme associated with this unit, VC1B *Friezes*. You are advised to watch this programme, and to work through the activities in the *Video Notes* related to the programme, before you continue.

In the following table we indicate whether the symmetry group of each type includes: reflections in a vertical axis (v); reflections in a horizontal axis (h); glide reflections (g); or rotations (r). We use Y for Yes and N for No.

Type	v	h	g	r
1	N	N	N	N
2	Y	N	N	N
3	N	Y	Y	N
4	N	N	Y	N
5	N	N	N	Y
6	Y	Y	Y	Y
7	Y	N	Y	Y

Notice that each of the seven rows of the table contains a *unique* combination of Ys and Ns. Therefore, in order to determine the type of a given frieze, we just need to ask in turn the following four questions.

Does the group of symmetries contain:
- reflections in a vertical axis (v)?
- reflections in a horizontal axis (h)?
- glide reflections (g)?
- rotations (r)?

Then, comparing the answers with the above table, we can determine the type of the frieze.

Example 4.1

Consider the frieze in Figure 4.1.

KOKOKOKOK

Figure 4.1

For this frieze, we find:

$$\begin{array}{cccc} v & h & g & r \\ N & Y & Y & N \end{array}$$

The frieze is consequently of Type 3. ◆

Exercise 4.4

Find the type of each of the friezes in Figure 4.2.

MWMWMWMWM

(a)

QQQQQQQQQ

(b)

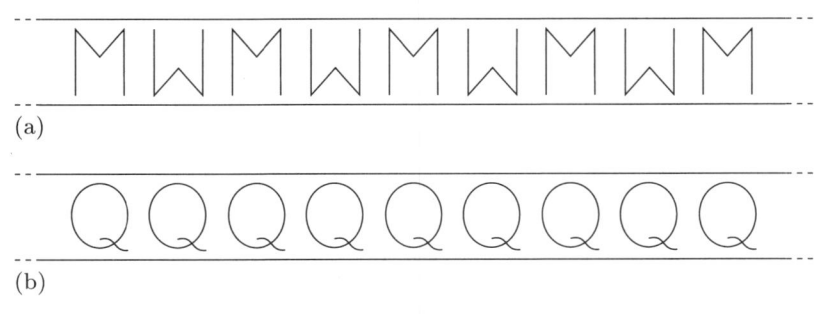

Figure 4.2

There is a certain amount of redundancy in the above procedure for finding the type of a frieze. We do not always have to ask each of the questions. Given some answers to questions, we can decide whether we already have enough information and, if not, which is the next suitable question to ask. In this way we arrive at a decision tree, shown in Figure 4.3, which will be the algorithm we use in future when dealing with friezes.

$v?$ Is there a reflection in a vertical axis?

$h?$ Is there a reflection in the horizontal axis?

$g?$ Is there a glide reflection?

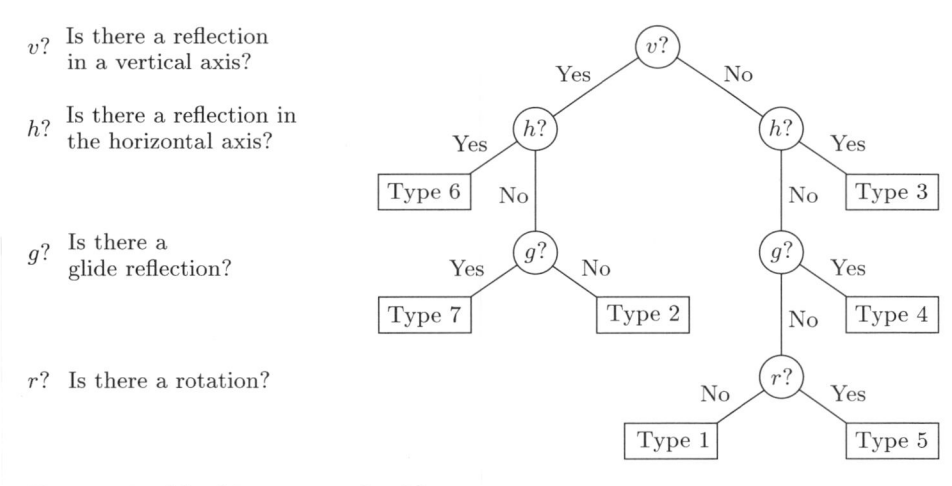

$r?$ Is there a rotation?

Figure 4.3 The frieze group algorithm.

Example 4.2

Consider the frieze in Figure 4.4.

NNNNNNNN

Figure 4.4

Using the Frieze Group Algorithm, the questions and answers we need are:

Is there a reflection in a vertical axis (v)?	– No.
Is there a reflection in the horizontal axis (h)?	– No.
Is there a glide reflection (g)?	– No.
Is there a rotation (r)?	– Yes.

The frieze is of Type 5. ♦

Exercise 4.5

Use the Frieze Group Algorithm to find the type of each of the friezes in Figure 4.5.

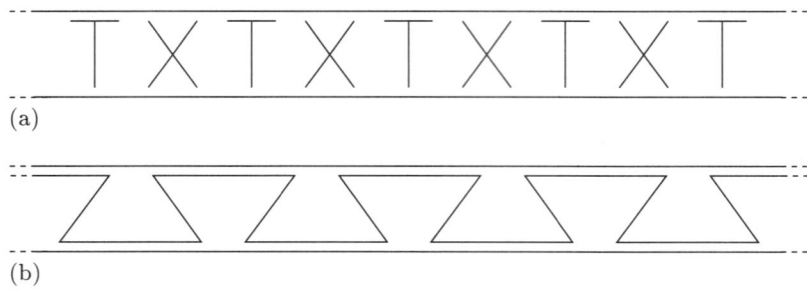

(a) TXTXTXTXT

(b) ZZZZZZ

Figure 4.5

We can use the Frieze Group Algorithm to introduce some notation for the frieze groups corresponding to the seven types. The frieze group given by a Type 1 frieze is found following a sequence of negative answers, and is written f_1. That is, we have

$$\Gamma(F_1) = T_1 = f_1.$$

The other types are found after one or two positive answers, and their groups are written as f with a suitable string of subscripts, where a subscript v (or h or g or r) corresponds to a positive answer to the question 'Is there a vertical reflection (or horizontal reflection or glide reflection or rotation)?'.

Example 4.3

A Type 2 frieze is found after a positive answer to 'Is there a vertical reflection?' and negative answers to 'Is there a horizontal reflection?' and 'Is there a glide reflection?'. The one positive answer results in the notation

$$\Gamma(F_2) = f_v.$$ ♦

Exercise 4.6

Use the above notation to describe the remaining frieze groups.

The above discussion, exercise and example have led to the following notation for the seven frieze groups:

$\Gamma(F_1) = f_1$
$\Gamma(F_2) = f_v$
$\Gamma(F_3) = f_h$
$\Gamma(F_4) = f_g$
$\Gamma(F_5) = f_r$
$\Gamma(F_6) = f_{vh}$
$\Gamma(F_7) = f_{vg}$

4.3 Only seven frieze groups

There is only one point that remains to be covered in our discussion of friezes. We have obtained an algorithm that enables us to determine the type of any given frieze, but we have not *proved* that the seven types of frieze are the only possibilities. We remedy that omission now.

> **Theorem 4.2 Seven frieze groups**
>
> There are precisely seven types of frieze, classified by the seven frieze groups f_1, f_v, f_h, f_g, f_r, f_{vh}, f_{vg}.

As we saw in Theorem 4.1, these fall algebraically into just four isomorphism classes.

Proof

What the Frieze Group Algorithm does for us is to classify friezes according to a set of answers. We have seen that the seven types of frieze described earlier have different frieze groups and are distinguished by the answers given in the algorithm. What this does not automatically rule out is that, given a set of answers, there might be another frieze with a new and different frieze group giving the same answers. We rule out this possibility by looking in detail at what the answers tell us for each of the seven types in turn. We shall show that, firstly, they allow us to construct all the symmetries in the appropriate group $\Gamma(F_i)$ in each case. We shall also show that, secondly, they allow us to infer that the symmetries in this group constitute all the symmetries of the frieze, because in each case we are able to show that, if there were any further symmetries, we would arrive at the contradictory position of being able to construct translations which are not multiples of the basic translation t (through one step) which generates T_1, the frieze's group of translational symmetries.

We here make use of the discussions of the Types 1–7 friezes in Sections 1–3.

Type 1 Here we have negative answers to there being vertical reflections, a horizontal reflection, glide reflections or rotations. All we are left with are translations, and, by the definition of a frieze, this group of translations must be T_1.

Type 2 Here we have a positive answer to there being vertical reflections, and negative answers to there being a horizontal reflection or glide reflections. From the translations t^n in T_1 and a vertical reflection v, we can construct all the symmetries in $\Gamma(F_2) = f_v$. All we are left with following our answers is the possibility of rotations and of vertical reflections other than those one step apart (i.e. those given by $t^n v$ ($n \in \mathbb{Z}$)). There cannot be rotations, as combining one of these with a vertical reflection would give a horizontal reflection or a glide reflection, which we know to be impossible. There cannot be reflections in other vertical axes, as combining these with those we already have would give new translations not in T_1.

Remember that
$\Gamma(F_2) = \{xy : x \in T_1, y \in Q_v;$
$vt = t^{-1}v\}$.

Type 3 Here we have a negative answer to there being vertical reflections, and a positive answer to there being a horizontal reflection. From the translations t^n in T_1 and the horizontal reflection h, we can construct all the symmetries in $\Gamma(F_3) = f_h$. All we are left with following our answers is the possibility of rotations and of glide reflections other than those through distances one step apart (i.e. those given by $t^n h$ ($n \in \mathbb{Z}$)). There cannot be rotations, as combining one of these with the horizontal reflection would give a vertical reflection, which we know to be impossible. There cannot be glide reflections through other distances, as combining these with those we already have would give new translations not in T_1.

Remember that
$\Gamma(F_3) = \{xy: \ x \in T_1, \ y \in Q_h; \ ht = th\}.$

Type 4 Here we have a negative answer to there being vertical reflections or a horizontal reflection, and a positive answer to there being glide reflections. If g is any such glide reflection, we find that g^2 is a translation. If g is chosen to give the smallest possible translation, then we find $g^2 = t$, that is, g glides through half a step. (If g^2 gave a longer translation than t then $(t^{-1}g)^2$ would give a smaller one, contradicting the choice of g.) From the translations in T_1 and this glide reflection g we can construct all the symmetries in $\Gamma(F_4) = f_g$. All we are left with following our answers is the possibility of rotations and of glide reflections other than those through distances one step apart (i.e. those given by $t^n g$ ($n \in \mathbb{Z}$)). There cannot be rotations, as combining one of these with a glide reflection would give a vertical reflection, which we know to be impossible. There cannot be glide reflections through other distances, as combining these with those we already have would give new translations not in T_1.

Remember that
$\Gamma(F_4) = \{x, \ xg: \ x \in T_1; \ g^2 = t, \ gt = tg\},$
where earlier we wrote $g_{1/2}$ rather than g for the glide reflection through half a step.

Type 5 Here we have negative answers to there being vertical reflections, a horizontal reflection or glide reflections, and a positive answer to there being rotations. From the translations t^n in T_1 and a rotation r we can construct all the symmetries in $\Gamma(F_5) = f_r$. All we are left with following our answers is the possibility of rotations other than those about points half a step apart (i.e. other than those given by $t^n r$ ($n \in \mathbb{Z}$)). There cannot be rotations about other points, since combining these with those we already have would give new translations not in T_1.

Remember that
$\Gamma(F_5) = \{xy: \ x \in T_1, \ y \in R_r; \ rt = t^{-1}r\}.$

Type 6 Here we have positive answers to there being vertical reflections and a horizontal reflection. From the translations t^n in T_1, a vertical reflection v and the horizontal reflection h, we can construct all the symmetries in $\Gamma(F_6) = f_{vh}$. All we are left with following our answers is the possibility of vertical reflections, glide reflections or rotations other than those given by $t^n v, t^n h$ or $t^n r$ ($n \in \mathbb{Z}$) respectively (where $r = hv = vh$). There cannot be any of these, as combining them with the corresponding symmetries which we already have would give new translations not in T_1.

Remember that
$\Gamma(F_6) = \{xy: \ x \in T_1, \ y \in V; \ ht = th, \ vt = t^{-1}v\}.$

Type 7 Here we have positive answers to there being vertical reflections and glide reflections and a negative answer to there being a horizontal reflection. As for Type 4, we can argue that there is a glide reflection g through half a step. From the translations t^n in T_1, a vertical reflection v and this glide reflection g, we can construct all the symmetries in $\Gamma(F_7) = f_{vg}$. All we are left with following our answers is the possibility of vertical reflections, glide reflections or rotations other than those given by $t^n v, t^n g$ or $t^n r$ ($n \in \mathbb{Z}$) respectively (where $r = gv$). There cannot be any of these, as combining them with the corresponding symmetries which we already have would give new translations not in T_1.

Remember that
$\Gamma(F_7) = \{xy, \ xgy: \ x \in T_1, \ y \in Q_v; \ vt = t^{-1}v, \ g^2 = t, \ gt = tg, \ vg = t^{-1}gv\},$
where earlier we wrote $g_{1/2}$ rather than g for the glide reflection through half a step and $r_{1/4}$ rather than r for the basic rotation.

Therefore, since all possible symmetries of a frieze (given in Theorem 1.2) are covered by the Algorithm, there are precisely seven types of frieze, classified by the seven frieze groups $f_1, \ f_v, \ f_h, \ f_g, \ f_r, \ f_{vh}, \ f_{vg}$. ∎

4.4 International notation

We have already introduced a notation for what we now know to be *all* seven frieze groups. Every text which deals with these groups introduces its own notation, and this has been no exception. Fortunately there is an internationally accepted notation. Below, we list our example of each frieze group along with our notation and the **International Notation**.

Example	Our Notation	International Notation
$\Gamma(F_1)$	f_1	$p111$
$\Gamma(F_2)$	f_v	$pm11$
$\Gamma(F_3)$	f_h	$p1m1$
$\Gamma(F_4)$	f_g	$p1a1$
$\Gamma(F_5)$	f_r	$p112$
$\Gamma(F_6)$	f_{vh}	$pmm2$
$\Gamma(F_7)$	f_{vg}	$pma2$

The explanation of the International Notation is as follows. The first symbol, p, has no particular significance for frieze groups. The second, third and fourth symbols refer in turn to reflections or glide reflections in vertical axes, to reflections in the horizontal axis and to the order of rotations. For the second and third symbols, if there is no reflection in that particular direction we write 1, if there is a reflection we write m and if there is no reflection but there is a glide reflection we write a. For the fourth symbol we just record the maximum order of a rotation: for friezes, any non-trivial rotation has order 2, and if no non-trivial rotation exists we write 1.

In *Unit GE4* you will see that, for the symmetry groups of wallpaper patterns, the first symbol can be either p or c, and so does carry a significance.

The International Notation for the symmetry groups of wallpaper patterns is constructed in a similar way. You will meet it in *Unit GE4*.

Exercise 4.7

Write out the result of Theorem 4.1 using the International Notation.

SOLUTIONS TO THE EXERCISES

Solution 1.1

We have
$$\begin{aligned} vr &= vvh && \text{(since } r = vh) \\ &= v^2 h \\ &= eh && \text{(since } v^2 = e) \\ &= h. \end{aligned}$$

Solution 1.2

(a) Composition gives
$$\begin{aligned} hvh &= hhv && \text{(since } vh = hv) \\ &= h^2 v \\ &= ev && \text{(since } h^2 = e) \\ &= v, \end{aligned}$$
which is in standard form.

(b) Since $h^2 = v^2 = e$, we have $h^{-1} = h$ and $v^{-1} = v$. Hence
$$\begin{aligned} (vh)^{-1} &= h^{-1} v^{-1} \\ &= hv \\ &= vh && \text{(since } hv = vh), \end{aligned}$$
which is in standard form.

Solution 1.3

To simplify
$$r^3 s r^4,$$
we need to interchange the s and second power of r. To do this we take the relation
$$srs^{-1} = r^{-1}$$
and raise it to the fourth power, giving
$$\left(srs^{-1}\right)^4 = \left(r^{-1}\right)^4,$$
which reduces to
$$\begin{aligned} sr^4 s^{-1} &= r^{-4} \\ &= r^2 && \text{(since } r^6 = e), \end{aligned}$$
and hence
$$sr^4 = r^2 s.$$
So
$$\begin{aligned} r^3 s r^4 &= r^3 r^2 s \\ &= r^5 s, \end{aligned}$$
which is in standard form (with $m = 5, n = 1$).

Solution 1.4

This isometry is described by the standard form

$t[\mathbf{p}]\, q[\pi/4],$

where $\mathbf{p}$ is the vector of length 2 in the direction defined by the position vector $(1,1)$. Thus $\mathbf{p} = (\sqrt{2}, \sqrt{2})$.

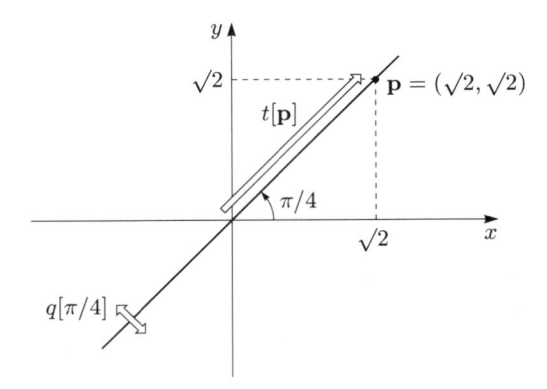

Using Equation 23 of the Isometry Toolkit, we obtain:

$$t[(\sqrt{2},\sqrt{2})]\, q[\pi/4] : (x,y) \mapsto (x\cos\pi/2 + y\sin\pi/2 + \sqrt{2}, x\sin\pi/2 - y\cos\pi/2 + \sqrt{2})$$
$$= (y + \sqrt{2}, x + \sqrt{2});$$

or alternatively, using Equation 23a, we obtain:

$$t[(\sqrt{2},\sqrt{2})]\, q[\pi/4] : \begin{bmatrix} x \\ y \end{bmatrix} \mapsto \begin{bmatrix} \cos\pi/2 & \sin\pi/2 \\ \sin\pi/2 & -\cos\pi/2 \end{bmatrix} \begin{bmatrix} x \\ y \end{bmatrix} + \begin{bmatrix} \sqrt{2} \\ \sqrt{2} \end{bmatrix}$$
$$= \begin{bmatrix} 0 & 1 \\ 1 & 0 \end{bmatrix} \begin{bmatrix} x \\ y \end{bmatrix} + \begin{bmatrix} \sqrt{2} \\ \sqrt{2} \end{bmatrix}.$$

Solution 1.5

The group of symmetries $\Gamma(O)$ consists of all those isometries that preserve the origin, and hence consists of all the orthogonal transformations, represented by all the corresponding orthogonal matrices.

The subgroup of direct symmetries $\Gamma^+(O)$ consists of all the rotations, i.e. all those represented by matrices of the form

$$\begin{bmatrix} \cos\theta & -\sin\theta \\ \sin\theta & \cos\theta \end{bmatrix}.$$

The subgroup of translations $\Delta(O)$ contains only the identity transformation.

Solution 1.6

$\Gamma(R_1) = \{e\}$
$\Gamma(R_2) = Q_v$
$\Gamma(R_3) = Q_h$
$\Gamma(R_4) = \{e\}$
$\Gamma(R_5) = R_r$
$\Gamma(R_6) = V$
$\Gamma(R_7) = Q_v$

Solution 1.7

(a) Here the group of translations is $\{e\}$. The half line is not mapped onto itself by any translation other than the identity. This is not a frieze.

(b) Here the group of translations is isomorphic to the group of real numbers, $\mathbb{R}$, under addition, since the parallel lines are mapped onto themselves by a translation through any distance. This is not a frieze.

(c) Here the group of translations is the group $T_1 = \langle t[\mathbf{a}] \rangle$, where $\mathbf{a}$ is as shown below. This is a frieze.

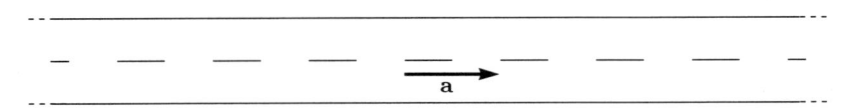

Solution 1.8

Any flag can be mapped to any other by a horizontal translation by the distance between the flags, and such a translation maps the entire frieze to itself and is therefore a symmetry. No flag can be mapped onto any other by an isometry other than a translation. Therefore, the only symmetries are horizontal translations though integer multiples of the distance between a pair of adjacent flags, and so

$$\Gamma(F_1) = T_1 = \langle t[\mathbf{a}] \rangle,$$

where $\mathbf{a}$ is the vector shown below.

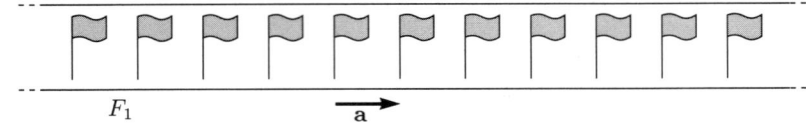

Furthermore, we also have

$$\Gamma^+(F_1) = \Delta(F_1) = T_1.$$

Solution 2.1

The symmetries are

$\quad t^n \quad$ and $\quad t^n v, \quad$ where n is any integer (positive or negative),

giving

$$\ldots, t^{-2}, t^{-1}, e, t, t^2, \ldots$$

and

$$\ldots, t^{-2}v, t^{-1}v, v, tv, t^2v, \ldots$$

Solution 2.2

Reflections are indirect symmetries, and the composite of two is a direct symmetry and hence a translation (there are no rotations in this case). To see which translation it is, we need only look at what it does to one point or, in this case, one vertical axis. The vertical axis of symmetry of our base rectangle is mapped by $v_{1/2}$ to the vertical axis of symmetry of the rectangle translated one place to the right. This in turn is mapped by $v_{-1/2}$ to the vertical axis of symmetry of the rectangle two steps to the left of the base rectangle. Hence we have

$$v_{-1/2}\, v_{1/2} = t^{-2}.$$

See Subsection 5.1 of *Unit IB1*.

In this diagram, $(a,0) = \mathbf{a}$, where $t = t[\mathbf{a}]$.

Alternatively, this may be done algebraically. Choose a coordinate system with the origin at the centre of the base rectangle and the x-axis pointing horizontally to the right along the frieze, as in the figure above. Then, for $t = t[\mathbf{a}]$, the reflection $v_{1/2}$ is $q[\frac{1}{2}\mathbf{a}, \pi/2]$, while $v_{-1/2}$ is $q[-\frac{1}{2}\mathbf{a}, \pi/2]$.

Now $\frac{1}{2}\mathbf{a}$ and $-\frac{1}{2}\mathbf{a}$ are each perpendicular to the reflection axis, so we can use Equation 12 of the Isometry Toolkit to express these as

$$v_{1/2} = t[\mathbf{a}]\, q[\pi/2], \quad v_{-1/2} = t[-\mathbf{a}]\, q[\pi/2].$$

Thus,

$$v_{-1/2}\, v_{1/2} = t[-\mathbf{a}]\, q[\pi/2]\, t[\mathbf{a}]\, q[\pi/2].$$

Since, with $v = q[\pi/2]$, we get $q[\pi/2](\mathbf{a}) = -\mathbf{a}$, we can use Equation 6b of the Toolkit to re-express the two middle terms on the right-hand side, $q[\pi/2]\, t[\mathbf{a}]$, as $t[-\mathbf{a}]\, q[\pi/2]$. Thus,

$$\begin{aligned} v_{-1/2}\, v_{1/2} &= t[-\mathbf{a}]\, t[-\mathbf{a}]\, q[\pi/2]\, q[\pi/2] \\ &= t[-\mathbf{a}]^2 \quad \text{(since } q[\pi/2]\, q[\pi/2] = e) \\ &= t^{-2} \quad \text{(since } t[\mathbf{a}] = t). \end{aligned}$$

Solution 2.3

$t^n = t[n\mathbf{a}] : (x, y) \mapsto (x + na, y)$.

Solution 2.4

From Equation 2.4 we have

$$\begin{aligned} v_{n+1/2} &= t[(2n+1)\mathbf{a}]\, q[\pi/2] \\ &= t[((2n+1)a, 0)]\, q[\pi/2]. \end{aligned}$$

Using Equation 23 of the Isometry Toolkit, this becomes

$$v_{n+1/2} : (x, y) \mapsto ((2n+1)a - x, y).$$

Alternatively, using Equation 23a of the Toolkit,

$$v_{n+1/2} : \begin{bmatrix} x \\ y \end{bmatrix} \mapsto \begin{bmatrix} -1 & 0 \\ 0 & 1 \end{bmatrix} \begin{bmatrix} x \\ y \end{bmatrix} + \begin{bmatrix} (2n+1)a \\ 0 \end{bmatrix}.$$

Solution 2.5

The elements of Q_v are e and v. The set

$$\{xy : x \in T_1,\ y = e\}$$

is just the set $\{t^n : n \in \mathbb{Z}\}$. Furthermore, by Equation 2.3, the set

$$\{xy : x = t^{2n},\ y = v,\ n \in \mathbb{Z}\}$$

is just the set $\{v_n : n \in \mathbb{Z}\}$, whereas, by Equation 2.4, the set

$$\{xy : x = t^{2n+1},\ y = v,\ n \in \mathbb{Z}\}$$

is just the set $\{v_{n+1/2} : n \in \mathbb{Z}\}$. Thus,

$$\{t^n, v_n, v_{n+1/2} : n \in \mathbb{Z}\} = \{xy : x \in T_1,\ y \in Q_v;\ vt = t^{-1}v\},$$

as required.

Solution 2.6

By Equation 2.3,

$$v_i = t^{2i} v,$$

and so the result of performing t^n followed by v_i is

$$\begin{aligned} v_i t^n &= t^{2i} v t^n \\ &= t^{2i} t^{-n} v \quad \text{(using the relation } vt = t^{-1}v \text{ and Theorem 1.1)} \\ &= t^{2i-n} v, \end{aligned}$$

which is in standard form.

Note that, by Equation 2.3, $t^{2i-n}v$ may also be written as $v_{i-n/2}$.

Solution 2.7

Using Equation 2.3, we have

$$v_j v_i = t^{2j} v t^{2i} v$$
$$= t^{2j} t^{-2i} v v \quad \left(\text{using } vt = t^{-1}v \text{ and Theorem 1.1}\right)$$
$$= t^{2(j-i)} \quad \left(\text{since } v^2 = e\right),$$

which is in standard form.

Note that, as we might have expected, we have obtained a translation.

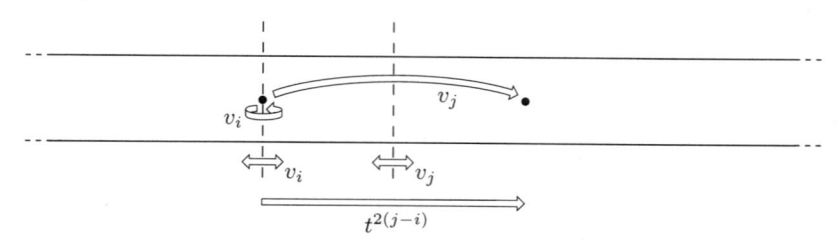

Solution 2.8

By the definition of g_i, we have

$$g_i t^n = t^i h t^n$$
$$= t^i t^n h \quad (\text{using the relation } ht = th \text{ and Theorem 1.1})$$
$$= t^{i+n} h,$$

which is in standard form.

Note that t^{i+n} may also be written as g_{i+n}.

Solution 2.9

By the definition of g_i and g_j, we have

$$g_j g_i = t^j h t^i h$$
$$= t^j t^i h h \quad (\text{using } ht = th \text{ and Theorem 1.1})$$
$$= t^{j+i} \quad \left(\text{since } h^2 = e\right),$$

which is in standard form.

Note that, as we might have expected, we have obtained a translation, which moves everything by the sum of the displacements of the two glide reflections.

Solution 2.10

We have, with $\mathbf{a} = (a, 0)$,

$$g_n = t^n h$$
$$= t[(na, 0)] \, q[0].$$

Using Equation 23 of the Isometry Toolkit card, we obtain

$$g_n : (x, y) \mapsto (x + na, -y),$$

or, using Equation 23a of the Toolkit,

$$g_n : \begin{bmatrix} x \\ y \end{bmatrix} \mapsto \begin{bmatrix} 1 & 0 \\ 0 & -1 \end{bmatrix} \begin{bmatrix} x \\ y \end{bmatrix} + \begin{bmatrix} na \\ 0 \end{bmatrix}.$$

Solution 2.11

$$g_{1/2} \, g_{1/2} = t_{1/2} \, h \, t_{1/2} \, h \quad (\text{since } g_{1/2} = t_{1/2} \, h)$$
$$= t_{1/2} \, t_{1/2} \, h \, h \quad (\text{since } h \, t_{1/2} = t_{1/2} \, h)$$
$$= t \, h^2 \quad (\text{since } t_{1/2} \, t_{1/2} = t)$$
$$= t \quad \left(\text{since } h^2 = e\right).$$

Solution 2.12

Multiplying both sides of the relation $g_{1/2}^2 = t$ by t^{-1} gives

$$t^{-1} g_{1/2}^2 = t^{-1} t = e.$$

From this it follows that

$$\left(t^{-1} g_{1/2}\right) g_{1/2} = e.$$

Since the product of $t^{-1} g_{1/2}$ and $g_{1/2}$ is the identity e, each is the inverse of the other, and so

$$g_{1/2}^{-1} = t^{-1} g_{1/2}.$$

To evaluate the more general inverse, we argue as follows.

$$\begin{aligned}
g_{n+1/2}^{-1} &= \left(t^n g_{1/2}\right)^{-1} && \text{(since } g_{n+1/2} = t^n g_{1/2}) \\
&= g_{1/2}^{-1} \left(t^n\right)^{-1} \\
&= t^{-1} g_{1/2} t^{-n} && \left(\text{since } g_{1/2}^{-1} = t^{-1} g_{1/2}\right) \\
&= t^{-1} t^{-n} g_{1/2} && \text{(using } g_{1/2} t = t g_{1/2} \text{ and Theorem 1.1)} \\
&= t^{-(n+1)} g_{1/2}.
\end{aligned}$$

Solution 3.1

The symmetries are

$$t^n \quad \text{and} \quad t^n r, \quad \text{where } n \text{ is any integer (positive or negative),}$$

giving

$$\ldots, t^{-2}, t^{-1}, e, t, t^2, \ldots$$

and

$$\ldots, t^{-2} r, t^{-1} r, r, tr, t^2 r, \ldots$$

Solution 3.2

We have

$$\begin{aligned}
r_{-1/2} r_{1/2} &= t^{-1} rtr && \text{(using the Frieze Card or the Isometry Toolkit)} \\
&= t^{-1} t^{-1} rr && \left(\text{using the relation } rt = t^{-1} r\right) \\
&= t^{-2} && \left(\text{since } r^2 = e\right).
\end{aligned}$$

Solution 3.3

From Equation 3.3,

$$r_n = t[(2na, 0)] \, r[\pi],$$

whose explicit form is

$$r_n : (x, y) \mapsto (2na - x, -y) \quad \text{(by Equation 22 of the Isometry Toolkit).}$$

Similarly,

$$r_{n+1/2} : (x, y) \mapsto ((2n+1)a - x, -y).$$

Alternatively, using Equation 22a of the Toolkit:

$$r_n : \begin{bmatrix} x \\ y \end{bmatrix} \mapsto \begin{bmatrix} -1 & 0 \\ 0 & -1 \end{bmatrix} \begin{bmatrix} x \\ y \end{bmatrix} + \begin{bmatrix} 2na \\ 0 \end{bmatrix};$$

$$r_{n+1/2} : \begin{bmatrix} x \\ y \end{bmatrix} \mapsto \begin{bmatrix} -1 & 0 \\ 0 & -1 \end{bmatrix} \begin{bmatrix} x \\ y \end{bmatrix} + \begin{bmatrix} (2n+1)a \\ 0 \end{bmatrix}.$$

Solution 3.4

The elements of R_r are e and r. The set
$$\{xy:\ x \in T_1,\ y = e\}$$
is just the set $\{t^n:\ n \in \mathbb{Z}\}$. Furthermore, by Equation 3.3, the set
$$\{xy:\ x = t^{2n},\ y = r,\ n \in \mathbb{Z}\}$$
is just the set $\{r_n:\ n \in \mathbb{Z}\}$, whereas, by Equation 3.4, the set
$$\{xy:\ x = t^{2n+1},\ y = r,\ n \in \mathbb{Z}\}$$
is just the set $\{r_{n+1/2}:\ n \in \mathbb{Z}\}$. Thus,
$$\{t^n, r_n, r_{n+1/2}:\ n \in \mathbb{Z}\} = \{xy:\ x \in T_1,\ y \in R_r,\ rt = t^{-1}r\},$$
as required.

Solution 3.5

We need to be slightly careful here. There are four possible types of flag: those facing upwards and flying to the right, those facing upwards and flying to the left, those facing downwards and flying to the right and those facing downwards and flying to the left. All four types are illustrated below.

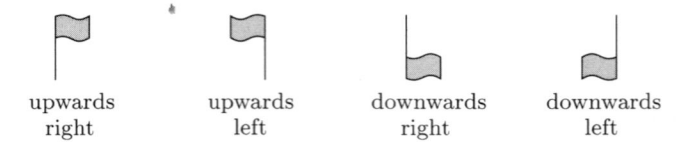

upwards right   upwards left   downwards right   downwards left

Now, only two of these types of flag appear in the frieze F_5: upwards right and downwards left. A reflected upwards right flag would become either an upwards left or a downwards right flag, and neither of these appear in the frieze. Similarly, a reflected downwards left flag would become either a downwards right flag or an upwards left flag, and neither of these appear in the frieze.

Therefore there are no reflection or glide reflection symmetries of the frieze. Hence, by Theorem 1.1,
$$\Gamma(F_5) = \{xy:\ x \in T_1,\ y \in R_r;\ rt = t^{-1}r\}.$$

Solution 3.6

$$\begin{aligned} r_i t^n &= t^{2i} r t^n & \text{(by Equation 3.3)} \\ &= t^{2i} t^{-n} r & \text{(using the relation } rt = t^{-1}r \text{ and Theorem 1.1)} \\ &= t^{2i-n} r, \end{aligned}$$

which is in standard form.

Note that, by Equation 3.3, $t^{2i-n} r$ may also be written as $r_{i-n/2}$.

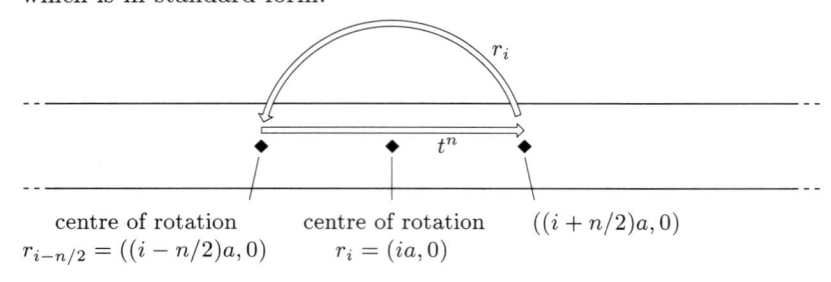

centre of rotation $r_{i-n/2} = ((i - n/2)a, 0)$ centre of rotation $r_i = (ia, 0)$ $((i + n/2)a, 0)$

The diagram shows that the centre of rotation $((i - n/2)a, 0)$ is fixed by the composite $r_i t^n = t^{2i-n} r = r_{i-n/2}$.

Solution 3.7

$$r_j r_i = t^{2j} r t^{2i} r \quad \text{(by Equation 3.3)}$$
$$= t^{2j} t^{-2i} r^2 \quad \text{(using } rt = t^{-1}r \text{ and Theorem 1.1)}$$
$$= t^{2(j-i)} \quad \text{(since } r^2 = e\text{),}$$

which is in standard form.

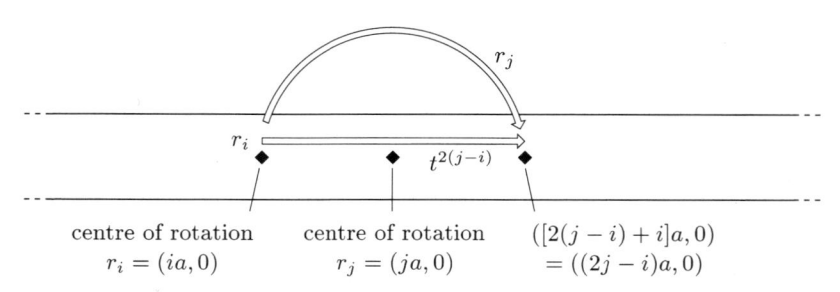

Note that, as we might have expected, we have obtained a translation.

centre of rotation $r_i = (ia, 0)$; centre of rotation $r_j = (ja, 0)$; $([2(j-i)+i]a, 0) = ((2j-i)a, 0)$

The diagram shows that the composite $r_j r_i$ is a translation that moves the centre of rotation for r_i a distance of $2(j-i)a$ to the right.

Solution 3.8

The direct symmetries are the translations and rotations, and so we have the group

$$\Gamma^+(F_6) = \{xy : x \in T_1, \ y \in R_r;\ rt = t^{-1}r\}.$$

We noted in Subsection 3.1 that this is a group and that the relation $rt = t^{-1}r$ holds.

Solution 3.9

$$r_j g_i = t^{2j} r t^i h \quad \text{(since } r_j = t^{2j}r \text{ and } g_i = t^i h\text{)}$$
$$= t^{2j} t^{-i} r h \quad \text{(using } rt = t^{-1}r \text{ and Theorem 1.1)}$$
$$= t^{2j-i} v \quad \text{(since } rh = v\text{),}$$

which is in standard form.

Note that, by Equation 2.3, $t^{2j-i}v$ may also be written as $v_{j-i/2}$.

Solution 3.10

We have:

$$v = q[\pi/2] : (x, y) \mapsto (-x, y) \quad \text{(by Equation 23 of the Isometry Toolkit);}$$
$$g_{1/2} = t\left[\left(\tfrac{1}{2}a, 0\right)\right] q[0] : (x, y) \mapsto \left(\tfrac{1}{2}a + x, -y\right) \quad \text{(by Equation 23 of the Toolkit).}$$

Thus

$$g_{1/2} v : (x, y) \mapsto \left(\tfrac{1}{2}a - x, -y\right).$$

Also

$$r_{1/4} = t\left[\left(\tfrac{1}{2}a, 0\right)\right] r[\pi] : (x, y) \mapsto \left(\tfrac{1}{2}a - x, -y\right) \quad \text{(by Equation 22 of the Toolkit).}$$

So $g_{1/2} v = r_{1/4}$.

Similarly,

$$v g_{1/2} : (x, y) \mapsto \left(-\left(\tfrac{1}{2}a + x\right), -y\right) = \left(-\tfrac{1}{2}a - x, -y\right).$$

Also

$$t^{-1} r_{1/4} = t[(-a, 0)]\, t\left[\left(\tfrac{1}{2}a, 0\right)\right] r[\pi] = t\left[\left(-\tfrac{1}{2}a, 0\right)\right] r[\pi] : (x, y) \mapsto \left(-\tfrac{1}{2}a - x, -y\right) \quad \text{(by Equation 22 of the Toolkit).}$$

So $v g_{1/2} = t^{-1} r_{1/4}$.

In terms of matrices and vectors, we have:

$$v : \begin{bmatrix} x \\ y \end{bmatrix} \mapsto \begin{bmatrix} -1 & 0 \\ 0 & 1 \end{bmatrix} \begin{bmatrix} x \\ y \end{bmatrix};$$

$$g_{1/2} : \begin{bmatrix} x \\ y \end{bmatrix} \mapsto \begin{bmatrix} 1 & 0 \\ 0 & -1 \end{bmatrix} \begin{bmatrix} x \\ y \end{bmatrix} + \begin{bmatrix} \tfrac{1}{2}a \\ 0 \end{bmatrix};$$

$$r_{1/4} : \begin{bmatrix} x \\ y \end{bmatrix} \mapsto \begin{bmatrix} -1 & 0 \\ 0 & -1 \end{bmatrix} \begin{bmatrix} x \\ y \end{bmatrix} + \begin{bmatrix} \tfrac{1}{2}a \\ 0 \end{bmatrix};$$

$$t^{-1} : \begin{bmatrix} x \\ y \end{bmatrix} \mapsto \begin{bmatrix} 1 & 0 \\ 0 & 1 \end{bmatrix} \begin{bmatrix} x \\ y \end{bmatrix} + \begin{bmatrix} -a \\ 0 \end{bmatrix}.$$

Therefore:

$$g_{1/2}v : \begin{bmatrix} x \\ y \end{bmatrix} \mapsto \begin{bmatrix} 1 & 0 \\ 0 & -1 \end{bmatrix} \left(\begin{bmatrix} -1 & 0 \\ 0 & 1 \end{bmatrix} \begin{bmatrix} x \\ y \end{bmatrix} \right) + \begin{bmatrix} \tfrac{1}{2}a \\ 0 \end{bmatrix} = \begin{bmatrix} -1 & 0 \\ 0 & -1 \end{bmatrix} \begin{bmatrix} x \\ y \end{bmatrix} + \begin{bmatrix} \tfrac{1}{2}a \\ 0 \end{bmatrix};$$

$$vg_{1/2} : \begin{bmatrix} x \\ y \end{bmatrix} \mapsto \begin{bmatrix} -1 & 0 \\ 0 & 1 \end{bmatrix} \left(\begin{bmatrix} 1 & 0 \\ 0 & -1 \end{bmatrix} \begin{bmatrix} x \\ y \end{bmatrix} + \begin{bmatrix} \tfrac{1}{2}a \\ 0 \end{bmatrix} \right) = \begin{bmatrix} -1 & 0 \\ 0 & -1 \end{bmatrix} \begin{bmatrix} x \\ y \end{bmatrix} + \begin{bmatrix} -\tfrac{1}{2}a \\ 0 \end{bmatrix};$$

$$t^{-1}r_{1/4} : \begin{bmatrix} x \\ y \end{bmatrix} \mapsto \begin{bmatrix} 1 & 0 \\ 0 & 1 \end{bmatrix} \left(\begin{bmatrix} -1 & 0 \\ 0 & -1 \end{bmatrix} \begin{bmatrix} x \\ y \end{bmatrix} + \begin{bmatrix} \tfrac{1}{2}a \\ 0 \end{bmatrix} \right) + \begin{bmatrix} -a \\ 0 \end{bmatrix} = \begin{bmatrix} -1 & 0 \\ 0 & -1 \end{bmatrix} \begin{bmatrix} x \\ y \end{bmatrix} + \begin{bmatrix} -\tfrac{1}{2}a \\ 0 \end{bmatrix}.$$

Hence $g_{1/2}v = r_{1/4}$ and $vg_{1/2} = t^{-1}r_{1/4}$.

Solution 3.11

$$\begin{aligned}
r_{j/2+1/4}\, g_{i+1/2} &= t^j\, r_{1/4}\, t^i\, g_{1/2} &&\text{(since } r_{j/2+1/4} = t^j r_{1/4} \text{ and } g_{i+1/2} = t^i g_{1/2}) \\
&= t^j\, g_{1/2}\, v\, t^i\, g_{1/2} &&\text{(since } g_{1/2}\, v = r_{1/4}) \\
&= t^j\, g_{1/2}\, t^{-i}\, v\, g_{1/2} &&\text{(using } vt = t^{-1}v \text{ and Theorem 1.1)} \\
&= t^j\, t^{-i}\, g_{1/2}\, v\, g_{1/2} &&\text{(using } g_{1/2}\, t = t\, g_{1/2} \text{ and Theorem 1.1)} \\
&= t^j\, t^{-i}\, g_{1/2}\, t^{-1}\, g_{1/2}\, v &&\text{(since } v\, g_{1/2} = t^{-1} g_{1/2}\, v) \\
&= t^j\, t^{-i}\, t^{-1}\, g_{1/2}\, g_{1/2}\, v &&\text{(using } g_{1/2}\, t = t\, g_{1/2} \text{ and Theorem 1.1)} \\
&= t^j\, t^{-i}\, t^{-1}\, t\, v &&\text{(since } g_{1/2}^2 = t) \\
&= t^{j-i}\, v,
\end{aligned}$$

Note that, by Equation 2.3, $t^{j-i}v$ may also be written as $v_{(j-i)/2}$.

which is in standard form.

Solution 4.1

The group $\Gamma(F_4)$ is isomorphic to $\Gamma(F_1)$. $\Gamma(F_4)$ is generated by $g_{1/2}$ and all its powers. Since $g_{1/2}^2 = t$, a general element in $\Gamma(F_4)$,

$$t^n \quad \text{or} \quad t^n g_{1/2},$$

can be written as

$$g_{1/2}^{2n} \quad \text{or} \quad g_{1/2}^{2n+1}.$$

Hence, identifying $g_{1/2}$ in $\Gamma(F_4)$ with t in $\Gamma(F_1)$ gives an isomorphism between these groups.

Solution 4.2

All pairs x, y of elements in each of $\Gamma(F_1)$, $\Gamma(F_3)$ and $\Gamma(F_4)$ *commute*, i.e. they satisfy the equality $xy = yx$, so these are Abelian.

In the remaining groups, either the relation $vt = t^{-1}v$ or the relation $rt = t^{-1}r$ shows that not all pairs of elements commute, and so they are not Abelian.

Solution 4.3

In $\Gamma(F_3)$ the element h is of order 2, i.e. $h^2 = e$, while in $\Gamma(F_1)$ there are no such elements.

Solution 4.4

(a) For this frieze, we find:

v	h	g	r
Y	N	Y	Y

The frieze is of Type 7.

(b) For this frieze, we find:

v	h	g	r
N	N	N	N

The frieze is of Type 1.

Solution 4.5

(a) The questions and answers are:

$v?$ — yes
$h?$ — no
$g?$ — no

Thus the frieze is of Type 2.

(b) The questions and answers are:

$v?$ — yes
$h?$ — no
$g?$ — yes

Thus the frieze is of Type 7.

Solution 4.6

For Type 3, the only positive answer is to 'Is there a horizontal reflection?'. Hence we use the notation f_h.

For Type 4, the only positive answer is to 'Is there a glide reflection?'. Hence we use the notation f_g.

For Type 5, the only positive answer is to 'Is there a rotation?'. Hence we use the notation f_r.

For Type 6 we have positive answers to 'Is there a vertical reflection?' and 'Is there a horizontal reflection?'. Hence we use the notation f_{vh}.

For Type 7, we have positive answers to 'Is there a vertical reflection?' and 'Is there a glide reflection?'. Hence we use the notation f_{vg}.

Solution 4.7

Abelian

$p111 \cong p1a1$
$p1m1$

Non-Abelian

$pm11 \cong p112 \cong pma2$
$pmm2$

OBJECTIVES

After you have studied this unit, you should be able to:

(a) explain how a group can be described in terms of elements expressed in standard form and a given set of relations between the generators;

(b) given a plane figure P, explain the concepts of the symmetry group $\Gamma(P)$, the direct symmetry group $\Gamma^+(P)$ and the translation group $\triangle(P)$;

(c) describe in general the symmetries that it is possible for a frieze to possess;

(d) recognize the seven different geometric types of symmetry group of a frieze, and describe each in terms of elements in standard form and a set of relations between generators;

(e) perform compositions of the elements of each of these types of group, when given in standard form;

(f) use the Frieze Group Algorithm to classify any given frieze and hence find the geometric type of its symmetry group;

(g) partition the geometric types of frieze group into isomorphism classes and say which classes are Abelian and which are non-Abelian;

(h) know the International Notation for the seven geometric types of frieze group.

INDEX

base rectangle 18
centre line of frieze 16
conjugacy relation 8
conjugate 8
direct symmetry 13
direct symmetry group 13
frieze 15
frieze group 16
Frieze Group Algorithm 39
frieze pattern 15
generator 7
group of symmetries of plain
    rectangular frieze 9
group of symmetries of rectangle 6
group of symmetries of regular hexagon
    8
International Notation for frieze groups
    43
isomorphism properties of frieze groups
    38
Klein group 6
perpendicular translation principle 20
plane figure 12
plane isometry 11
standard form 7, 8, 10, 22, 23, 26, 29,
    31, 36
symmetries of frieze 17
symmetry 12
symmetry group 12
translate 19
translational symmetries 13
translation group 13
Type 1 frieze 16
Type 2 frieze 18
Type 3 frieze 22
Type 4 frieze 24
Type 5 frieze 26
Type 6 frieze 30
Type 7 frieze 31
types of frieze group 41